Brain Tumors

This book offers a comprehensive exploration of brain tumors, beginning with a foundational understanding of their pathophysiology and extending through the latest advancements in diagnosis and treatment. The chapters examine the underlying mechanisms that drive the development and progression of brain tumors and present a detailed analysis of the incidence, distribution, and potential risk factors associated with brain tumors. The book explores the intricate relationship between brain tumors and visual disorders, and reviews the critical role of advanced imaging technologies in diagnosing brain tumors, evaluating the strengths and limitations of various modalities such as MRI and PET scans. Additionally, the book evaluates the effectiveness and precision of radiosurgery in targeting brain tumors, discussing its benefits and challenges in the context of non-invasive cancer treatment. The chapters also introduce us to antisense oligonucleotides as a novel therapeutic strategy, outlining their mechanism of action and potential to silence cancer-driving genes. The book highlights the latest developments in nanotechnology for brain tumor diagnosis, emphasizing the use of nanomaterials to enhance imaging quality and specificity.

Key Features:

- Provides a thorough exploration of brain tumors, from basic pathophysiology to advanced treatment options.
- Presents detailed analyses of the incidence, distribution, and potential risk factors associated with brain tumors.
- Discusses the critical role of imaging technologies, including MRI and PET scans, in the diagnosis of brain tumors.
- Introduces novel therapeutic strategies such as radiosurgery and antisense oligonucleotides for targeting brain tumors.
- Examines strategies to overcome the blood–brain barrier for effective drug delivery.

Toward the end, the book covers challenges of and strategies for delivering therapeutic agents across the blood–brain barrier and discusses phytochemicals in brain tumor treatment. This book serves as a useful source for researchers and students in oncology and neuroscience.

Brain Tumors

Advancements in Diagnostics and Innovative Therapies

Edited by
Xiaodong Li, Yawen Ma,
Zirong Fan, and Rekha Khandia

CRC Press
Taylor & Francis Group
Boca Raton London New York

CRC Press is an imprint of the
Taylor & Francis Group, an **informa** business

Designed cover image: Shutterstock

First edition published 2025
by CRC Press
2385 NW Executive Center Drive, Suite 320, Boca Raton FL 33431

and by CRC Press
4 Park Square, Milton Park, Abingdon, Oxon, OX14 4RN

CRC Press is an imprint of Taylor & Francis Group, LLC

ISBN: 9781032857572 (hbk)
ISBN: 9781032857602 (pbk)
ISBN: 9781003519706 (ebk)

DOI: 10.1201/9781003519706

Typeset in Times
by Apex CoVantage, LLC

Contents

Preface

Understanding brain tumors, among the most formidable challenges in contemporary medicine, has never been more critical. The rising incidence and associated mortality of brain tumors worldwide demand comprehensive insights into their complexities. This book aims to unravel the intricate tapestry of brain tumors, delving into their origins, diagnostic challenges, treatment strategies, and promising innovations poised to reshape brain cancer care.

Brain tumors, originating from the brain and central nervous system, pose significant health challenges. They affect individuals across various age groups, presenting complex diagnostic, treatment, and surgical hurdles. This book provides an in-depth analysis of the global prevalence of brain tumors, exploring the multifaceted factors contributing to their rise, such as chemical carcinogens, radiation exposure, and genetic predispositions.

The book discusses the pathophysiology of brain cancer, providing a detailed understanding of the disease, including its characterization, types, and clinical manifestations. The significant morbidity caused by brain tumors worldwide, affecting around 250,000 people annually, is highlighted. It reviews the epidemiology of brain cancer, examining regional and temporal variations in incidence rates, and explores potential risk factors, including genetic variables and environmental influences. The need for further research to understand the link between genes and the environment is emphasized.

The focus then shifts to neuro-ophthalmology, a subspecialty addressing visual issues arising from brain disorders. The role of the optic nerves in transmitting signals from the eyes to the brain, leading to various vision-related challenges, is highlighted. The book reviews imaging techniques in brain cancer analysis, discussing the importance of modalities such as computed tomography, magnetic resonance imaging, and advanced techniques like hyperstereoscopy and functional MRI. The potential of machine learning algorithms in enhancing diagnostic precision is underscored.

The book delves into stereotactic radiosurgery, a groundbreaking innovation in radiation therapy that enables precise delivery of therapeutic doses to target volumes while sparing adjacent healthy tissues. It explores antisense therapy for brain cancer treatment, highlighting the specificity of antisense oligonucleotides in targeting cancer cells. The pharmacology, mechanisms of action, and delivery methods for antisense drugs are discussed.

Nanomaterials' role in diagnosing brain tumors is examined, with a focus on their properties, their role in magnetic resonance and fluorescence imaging, and their utility in liquid biopsy specimens. The connection between neurodegeneration, particularly Alzheimer's disease, and viral infections is explored. The association of viruses like HSV1, Epstein-Barr virus, and COVID-19 with Alzheimer's and antiviral interventions is reviewed.

The introduction of green nanomedicines is emphasized, highlighting their environmental sustainability and tissue-specific targeting in brain tumor therapy. Various methodologies for synthesizing nanoparticles through biomaterial-derived methods are reviewed. Drug delivery mechanisms to brain tumor sites are discussed, highlighting pathways such as nasal transportation, cerebrospinal fluid pathways, and the role of nanomaterials in drug delivery.

Finally, the potential of phytochemicals in brain tumor therapy is explored. The challenges of the blood–brain barrier, off-target delivery, and cytotoxicity to normal cells are discussed, and various phytochemicals that can inhibit brain tumor development are reviewed. This book is an indispensable resource for medical professionals, researchers, and anyone seeking a profound understanding of brain tumors. It offers a comprehensive exploration from their inception to cutting-edge therapies, providing valuable insights into the complexities and innovations in brain cancer care.

Editors

Dr. Xiaodong Li is a Chief Physician at Shengjing Hospital of China Medical University, 36 Sanhao St, Heping District, Shenyang, Liaoning, China with an MDPHD in neurosurgery. With over 13 years of experience and having completed more than 1,500 neurosurgical operations, Dr. Li specializes in epilepsy surgery and cerebrovascular disease, demonstrating profound expertise in cerebral vascular bypass surgery and the treatment of cerebral aneurysms, among other conditions.

Dr. Yawen Ma, also a Chief Physician and Neurosurgeon at Shengjing Hospital of China Medical University, 36 Sanhao St, Heping District, Shenyang, Liaoning, China, holds an MDPHD. Dr. Ma has a distinguished background in treating intracranial diseases, including tumors and vascular diseases, since 2017.

Dr. Zirong Fan is working as a Lecturer and Researcher at Shengjing Hospital of China Medical University, 36 Sanhao St, Heping District, Shenyang, Liaoning, China and holds an MDPHD in neurosurgery. Graduating in 2020, Dr. Fan is working toward understanding the molecular mechanism of brain tumors and innovative strategies for diagnosis and treatment of brain tumors.

Dr. Rekha Khandia serves as an Assistant Professor in the Department of Biochemistry and Genetics at Barkatullah Vishwavidyalaya, Bhopal, Hoshangabad Road, Bhopal, Madhya Pradesh, India. Dr. Khandia's work stands at the intersection of molecular diagnostics, vaccine development, neurosciences, brain tumors, and the study of infectious diseases through the lens of genetics and biochemistry. She has published more than 100 research articles in respected journals and authored two books. She is also actively involved in reviewing high-impact journals and participating in various academic forums.

Contributors

Mansi Agarwal
Department of Biotechnology, University of
Engineering & Management
Kolkata, West Bengal, India

Nivedita Chatterjee
Department of Biotechnology, University of
Engineering & Management
Kolkata, West Bengal, India

Sanchari Das
Department of Biotechnology, University of
Engineering & Management
Kolkata, West Bengal, India

Zirong Fan
Department of Neurosurgery, Shengjing
Hospital of China Medical University
Shenyang, Liaoning, China

Sayantani Garai
Department of Biotechnology, University of
Engineering & Management
Kolkata, West Bengal, India

Pankaj Gurjar
Department of Science and Engineering, Novel
Global Community Educational Foundation
Hebersham, NSW, Australia

Rekha Khandia
Department of Biochemistry and Genetics
Barkatullah University
Bhopal, MP, India

Dibyajit Lahiri
Department of Biotechnology, University of
Engineering & Management
Kolkata, West Bengal, India

Xiaodong Li
Department of Neurosurgery, Shengjing
Hospital of China Medical University
Shenyang, Liaoning, China

Yawen Ma
Department of Neurosurgery, Shengjing
Hospital of China Medical University
Shenyang, Liaoning, China

Dipro Mukherjee
Department of Biotechnology, University of
Engineering & Management
Kolkata, West Bengal, India

Moupriya Nag
Department of Biotechnology, University of
Engineering & Management
Kolkata, West Bengal, India

1 Pathophysiology of Brain Cancer

Xiaodong Li[1]*, Sayantani Garai[2], Dipro Mukherjee[2], Dibyajit Lahiri[2], and Moupriya Nag[2]
[1]Department of Neurosurgery, Shengjing Hospital of China Medical University, Shenyang, Liaoning, China
[2]Department of Biotechnology, University of Engineering & Management, Kolkata, West Bengal, India
*Corresponding author

1.1 INTRODUCTION

Brain cancer or brain tumors develop when cells grow abnormally or uncontrollably, forming redundant tissue masses [1]. Essentially, they represent cellular overgrowth resulting in tumor formation, which can be benign (non-cancerous) or malignant (cancerous) [2]. Although brain cancer is less common than other cancer types, it significantly contributes to morbidity across various age groups, often affecting children [3]. Each year, approximately 250,000 people worldwide are diagnosed with primary brain tumors [4], making them the second most common cancer in children under 15, following lymphoblastic leukemia [5]. In 2012, brain tumors were the ninth most common type of cancer and the eighth leading cause of cancer death in the United Kingdom [6]. In the United States, brain cancer causes an estimated 13,000 deaths annually [7]. While studies suggest lower brain cancer incidence in less developed nations, this may be due to inadequate diagnosis resulting from poor healthcare facilities rather than an actual lower incidence of the disease [6].

The survival rate for brain cancer is very low due to poor prognosis and the extreme severity of the disease. Diagnosing brain cancer involves techniques that are often costly and invasive, such as tissue biopsy. Diagnosis can be achieved through magnetic resonance imaging (MRI), computed tomography (CT) scans, tissue biopsies, biomarker tests, and close monitoring of tumor-associated symptoms. Despite advancements in medical technologies, including machine learning approaches in image analysis [7, 8], MRI and CT scans followed by manual analysis remain the most common and widely accepted methods for cancer detection [9]. Even after successful diagnosis and treatment, the survival rate for patients with malignant tumors is typically less than five years.

Treatment options for brain cancer patients depend on the tumor's location, grade or malignancy, patient age, and their pathological and functional condition. Radiotherapy, which uses X-rays, beta, or gamma radiation to kill tumor cells while sparing normally functioning brain cells, is the most commonly used treatment. Chemotherapy, however, is not always suitable because the blood–brain barrier (BBB) prevents certain drugs from reaching the tumor. Surgical resection can be a viable option for complete or partial removal of tumor cells in some cases of meningiomas, gliomas, and most pituitary adenomas. However, surgery is not viable for tumors in tricky locations or in cases of inoperable craniotomy, where radiotherapy or chemotherapy is preferred.

Despite significant research advancements in brain carcinoma and tumor treatments, the life expectancy of brain cancer or neuro-oncological patients remains limited to a few years [10]. This highlights the need for more in-depth research into the pathophysiology of brain cancer and the development of new drug-repurposing strategies, technologies, and diagnostic methods for early prognosis and improved treatment.

DOI: 10.1201/9781003519706-1

1.2   CHARACTERIZATION OF BRAIN TUMORS

Brain tumors can be of two kinds: malignant and non-malignant. Malignancy refers to the condition where cells divide abnormally, without control, and invade nearby cells or tissues to behave similarly [11]. Therefore, all brain cancers can be referred to as brain tumors, but not all brain tumors are necessarily cancerous. Even though some tumors are benign, or non-cancerous, they can still pose a major threat to the patient and be dangerous depending on their location and size. In contrast, malignant or cancerous brain tumors grow rapidly and affect the surrounding vital structures in the brain, leading to severe risk factors and poor survival rates post-diagnosis [12, 13].

Brain tumors, whether malignant or not, may originate in the brain itself or spread to the brain from other parts of the body, with the latter being more common. Tumors that originate within the brain are categorized as primary brain tumors, examples of which include meningioma, glioma, glioblastoma, and pituitary tumors. Malignant tumors originating elsewhere in the body, such as the breast, lung, skin, or colon, can spread through metastasis and infect the brain cells, forming secondary or metastatic tumors. Metastatic tumors are more common and are associated with higher risk factors and an elevated capacity for growth and invasion [14].

There are approximately 120 to 130 types of brain tumors characterized by location, origin, severity, and the age groups they affect. The most common malignant brain tumors are gliomas and meningiomas. Gliomas account for about 30% of all brain tumors (both benign and malignant) and approximately 80% of malignant ones [15, 16]. They represent a broad category of primary brain tumors originating in the neuroglial cells surrounding the neurons of the brain and spinal cord. These tumors can be histologically classified as astrocytomas, ependymomas, glioblastomas (WHO grade IV), oligodendrogliomas, and so on. Further characterization is based on location, differentiation pattern, and the presence or absence of anaplastic features.

Among the several types, glioblastoma multiforme (GBM) is one of the most dangerous, invasive, and aggressive brain tumors [17–19]. GBM presents extremely low chances of survival post-treatment due to the highly invasive nature of the cancer cells, which infiltrate adjacent brain cells and tissues, leading to rapid spread and metastasis. Cells in GBM are often characterized by diffuse infiltration, a high rate of mitosis, cellular pleomorphism [20], and standard drug resistance properties, necessitating an extremely aggressive treatment approach [21].

Meningiomas are another group of brain-associated tumors, constituting approximately 20% of primary intracranial tumors [22]. Most meningiomas are benign (WHO grade I), but around 5–15% of cases are atypical (WHO grade II), and 1–3% are anaplastic or malignant [23]. Other brain tumors, such as primary central nervous system (CNS) lymphoma, acoustic neuroma, and pineoblastoma, occur less frequently and are considered rare tumors. A brief overview of the types and their associated clinicopathological features is summarized in Table 1.1 and is depicted in Figure 1.1.

1.3   PATHOPHYSIOLOGY AND TUMOR MICROENVIRONMENT

Tumor cells function differently from normal, unaffected cells and can be characterized by unique molecular signatures. Brain tumors are not simply clusters of abnormal cells growing uncontrollably in a homogeneous fashion; instead, they resemble an organ system formed by subpopulations of cells with specific compartmentalizations, including cellular and extracellular components, inflammatory cells, and their own vasculatures [35]. Tumors consist of a mixture of cancerous cells at different stages and non-cancerous stromal cells that constitute the tumor microenvironment (TME) (Figure 1.2). The cancerous cells progress through various stages, namely hyperplasia, metaplasia, anaplasia, dysplasia, and neoplasia.

In hyperplasia, cells begin to show abnormalities, though they appear normal under microscopic analysis. Abnormalities become more apparent in the metaplasia stage and become increasingly morphologically distinguishable in subsequent stages. In anaplasia, cells start invading surrounding

TABLE 1.1

Different Types of Brain Tumors, Their Location of Development, and Associated Risks and Symptoms

Tumor Type	Location of Development	Affected Areas or Symptoms	Malignancy or Risk	Occurrence	Ref.
Primary central nervous system lymphoma	Lymphocytes in the brain, eyes, spinal cord, or the membranes covering the brain and spinal cord	Abnormal B-cell production, seizures, balancing difficulties, blurred vision	High mortality rate as surgery is not possible	4 in every 100 brain or spinal cord tumors	[24]
Pineal region tumors	Cells in and around the pineal gland	Excessive production of pituitary hormones, blockage of cerebrospinal fluid channels	Malignancy in grade IV pineoblastoma	~1%	[25]
Meningioma	Membranes that surround the brain and spinal cord	Cause seizures, hearing and eyesight loss depending on the location of the tumor	Malignant in high-grade (grade 3) meningiomas	Around 21 of 100 brain tumors	[26]
Acoustic neuroma (vestibular schwannoma)	Schwann cells, which wrap around the peripheral vestibulocochlear nerve	Cause hearing and balance problems, neurological problems, dizziness, or vertigo	Benign, usually does not spread to other regions of the body, slow growth	6 out of 100 brain tumors	[27]
Glioblastoma multiforme (GBM)	Astrocytes that support nerve cells	Highly invasive and aggressive; symptoms include seizures, headaches, and fever	Lethal with high malignancy	1 per 10,000 cases	[15, 28]
Astrocytoma (children's cancer)	Astrocytes	Seizures, double vision, increasing head circumference in infants	Most common glioma in children; malignant	~40% of brain and spinal cord tumor in children	[29]
Pituitary tumors	Pituitary gland	Press on the nearby optic nerve hampering eyesight; overproduction or abnormal production of hormones	Very few are malignant	~8% brain tumors	[30]
Oligodendroglioma	Glial cells called oligodendrocytes	Headaches and seizures	Grade III oligodendrogliomas are malignant	~3% of brain tumors	[31, 32]
Craniopharyngioma	Base of brain, above the pituitary gland	Changes in hormone levels, headaches, pressure build-up in brain	benign brain tumors	6–13% in children and 1–3% in adults	[33]
Haemangioblastoma	Cells that line the blood vessels in the brain, spinal cord, and brain stem	Hamper body balance and speech coordination; build-up of cerebrospinal fluid	Benign tumor; rarely spread to other regions	~2% of brain tumors	[34]

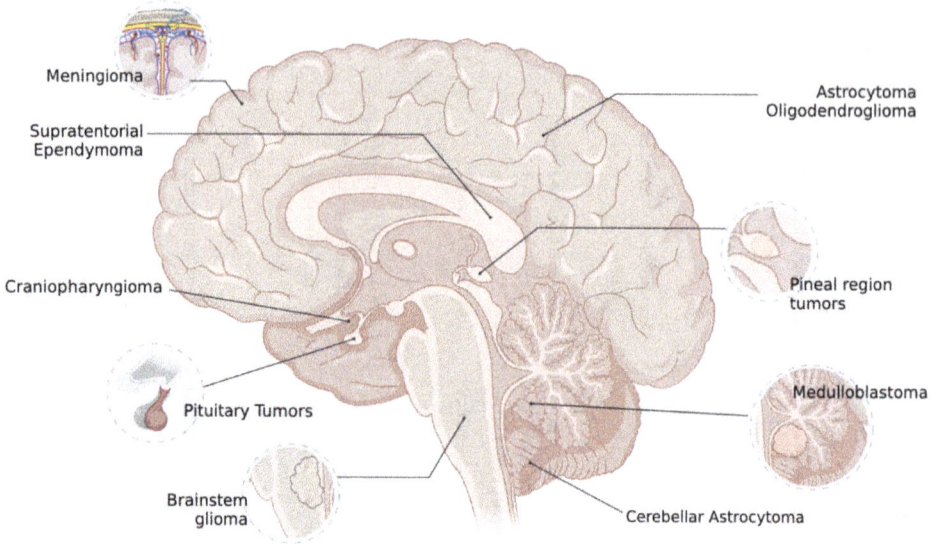

FIGURE 1.1 Anatomical locations and types of brain tumors. This diagram illustrates various types of brain tumors and their respective locations within the brain.

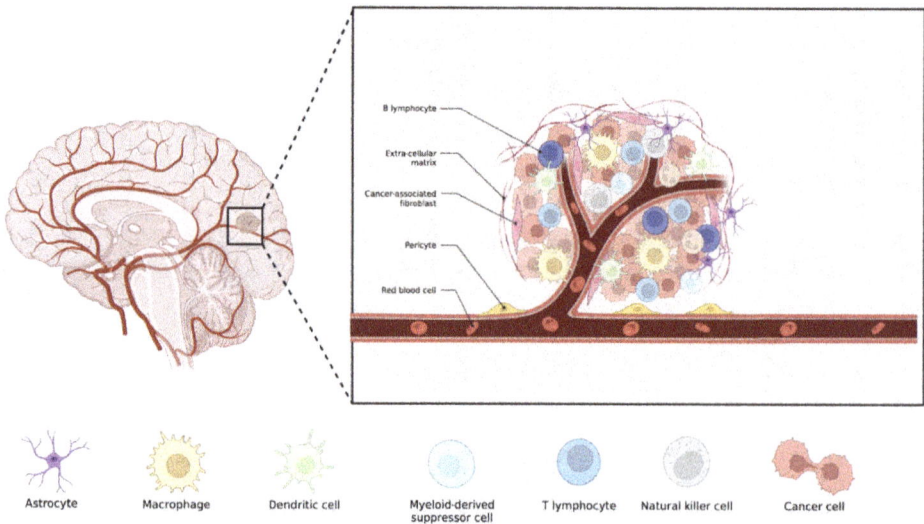

FIGURE 1.2 The tumor microenvironment in the brain. The left side shows a schematic of the brain highlighting the region of interest. The right side zooms into the tumor microenvironment, detailing the interactions between various cell types and structures within the tumor.

tissues and can even enter the bloodstream, leading to metastasis. Non-cancerous cells in the TME include pericytes, fibroblasts, endothelial cells, immune cells, inflammatory cells, and cells of myeloid origin [36].

Besides cellular components, the TME features differences from the normal physiology of brain tissue, such as variations in the composition of the extracellular matrix, increased vasculature, and the presence of growth factors.

1.4 CAUSES AND TRIGGERS OF CANCER DEVELOPMENT

Carcinogens are chemicals, physicochemical substances, irradiations, and environmental conditions that can trigger mutations in genes. The accumulation of such carcinogen-induced mutations results in improper functioning of natural cell lifecycles and uncontrolled proliferation, leading to tumor formation and malignancy. The exact causes of brain cancer and what triggers a normally functioning cell to become cancerous are poorly understood, as it arises from a complex interplay of multiple factors [37]. More often than not, cancer is triggered by a combination of multiple agents rather than a single root cause.

While the causal factors of cancers like lung cancer can be directly linked to tobacco use and smoking and kidney cancer to common carcinogenic solvents, the causes of brain cancer are more complex. Environmental and occupational causes of brain cancer can be linked to exposure to ionizing and non-ionizing radiation from mobile and cellular devices, exposure to pesticides, and chemical carcinogens such as N-nitroso compounds, vinyl chlorides, and hair dyes. Genetic contributors, like mutations and deletions in tumor suppressor genes such as P53, can also cause certain brain tumors [38].

Studies suggest that individuals with conditions such as neurofibromatosis type 2, endocrine neoplasia, and celiac disease are more prone to developing brain cancer [39, 40]. Recent findings indicate that heavy mobile phone usage for extended cumulative hours is associated with higher risks of glioma [41]. Additionally, the long-term consumption of aspartame, an artificial sweetener, is known to cause degenerative changes in the myelin fibers of nervous tissue, increasing the risk of CNS tumor development [42].

1.5 BRAIN STEM CELLS IN TUMOR FORMATION

The role of brain stem cells in the development and progression of cancer is significant because most mature brain or CNS cells are differentiated and lack the ability to replicate or divide [43]. Stem cells, on the contrary, can replicate and usually have a long lifespan, increasing the likelihood of accumulating multiple mutations over time, which can eventually lead to cancer. Moreover, stem cells possess the property of "self-renewal," enabling them to replicate or generate new stem cells identical to the undifferentiated parental ones. This property allows stem cells to produce similarly proliferating daughter stem cells with the same differentiation capabilities, whereas progenitor cells become progressively differentiated with each round of replication and eventually cease to proliferate [44].

This suggests that brain cancer likely originates from these self-renewing stem cells or from progenitor cells in which the mechanism for progressive differentiation is somehow turned off [45, 46].

1.6 CLINICAL MANIFESTATIONS

The symptoms of brain cancer often show minimal correlation with malignancy, meaning that both cancerous and non-cancerous tumors can produce similar symptoms [47]. As tumor cells grow, they exert mechanical effects on the surrounding tissue and microvasculature, leading to various clinical manifestations such as headaches and edema. The clinical manifestations of brain tumors mainly differ based on their location or the area of brain tissue they affect. The symptoms of common types of brain tumors are detailed in Table 1.1.

Headaches and seizures are common to most brain tumors. Headaches primarily occur due to the build-up of pressure in the intracranial space and are considered early signs of cancer [48]. While headaches can also be associated with stress, migraines, and other clinical conditions, certain characteristics—such as progressive worsening of the headache and headaches that cause awakening from sleep—are more specific to cancer [48].

Site-specific symptoms of brain tumors include personality changes, loss of balance, hearing loss, worsening of memory and reasoning ability, poor vision, endocrine defects, and more. Behavioral changes are often observed in patients where the tumor damages the frontal, temporal, and parietal lobes of the brain, as these regions are associated with emotions, mood, behavior, and judgment [49, 50].

1.7   CANCER PROGRESSION AND MALIGNANCY

Anaplastic features such as mitotic activity, cellular proliferation rate, microvascular progression, and necrosis are used to assign malignancy grades, indicating the severity of cancer and the malignant behavior of tumors. A malignancy grade I indicates the least malignant, while WHO grade IV represents the most malignant stage. The transformation or progression of cells through malignancy grades depends on various physiological factors.

The TME significantly influences the progression, invasiveness, and aggression of the tumor [51]. Recent studies have shown that microenvironmental stress, such as the presence of reactive radicals, pH variations, and hypoxia, can induce the differentiation of tumor cells [52]. Treating both primary brain tumors and metastatic tumors is challenging because systemic treatments are obstructed by the complications of the BBB [53, 54], and even surgical removal or radiation treatments have a median survival rate of less than 30% [55, 56].

In addition to the BBB and TME, other factors contributing to the development and progression of brain tumors include neuroinflammation, ion pump and protein channel deregulation [57], and distortions in cell death pathways. Deactivation or deregulation of proteins associated with signal transduction and the natural death of cells has been frequently observed in glioma [58]. Defects and hindrances in the cascade leading to cell death or apoptosis are major promoting causes of cancer progression, proliferation, invasiveness, and drug resistance, underscoring the need for repurposing available drugs and seeking alternative or synergistic therapeutics [3, 59].

1.8   CONCLUSION

Brain cancer remains a fatal disease with very limited non-surgical and non-radiative treatment options and extremely low survival rates. Despite extensive research worldwide, our understanding of the exact causal factors of brain cancer and its pathophysiology is still limited. Cancer prognosis remains a significant challenge, necessitating the development of better diagnostic and treatment methods.

The stringent and harsh treatment procedures, such as radiation and chemotherapy, induce numerous side effects and cause substantial damage to other organs. Surgery is also notoriously difficult due to complications related to the BBB. Although cutting-edge technologies in nanomedicine, drug repurposing, personalized drugs, and machine learning approaches in imaging analysis offer promise, the current medical landscape of cancer research has yet to reach its full potential.

REFERENCES

[1] Brain cancer: Causes, types & symptoms. *Healthline.* www.healthline.com/health/brain-cancer#overview

[2] Kaifi, R. (2023). A review of recent advances in brain tumor diagnosis based on AI-based classification. *Diagnostics, 13,* 3007.

[3] Pasi, F., Persico, M. G., Marenco, M., Vigorito, M., Facoetti, A., Hodolic, M., . . . Aprile, C. (2020). Effects of photons irradiation on 18F-FET and 18F-DOPA uptake by T98G glioblastoma cells. *Frontiers in Neuroscience, 14,* 589924.

[4] World Health Organization (2014). *Chapter 5.16. World cancer report 2014.* http://piblications.iarc.fr/ Non-Series-Publication/World Cancer-Report-2014

[5] McGuire, S. (2016). World cancer report 2014. *Advances in Nutrition, 7*(2), 418–419. Geneva, Switzerland: World Health Organization, International Agency for Research on Cancer, WHO Press, 2015.

[6] Cancer Research UK. *Brain, other CNS and intracranial tumors statistics.* Retrieved October 27, 2014, from www.cancerresearchuk.org/cancer.info/csancerstats/types/brain/

[7] Greenlee, R. T., Murray, T., Bolden, S., et al. (2000). Cancer statistics. *CA: A Cancer Journal for Clinicians, 50*(1), 7–33.

[8] *General information about adult brain tumors.* www.cancer.gov.cancertopics/pdq/treatment/adultbrain/ Patient/page1/All pages

[9] *Adult brain tumors treatment.* (www.cancer.gov/cancertopics/pdq/treatmentadultbrain/HealthProfessional/ page1/Allpages).

[10] Aldape, K., Brindle, K. M., Chesler, L., Chopra, R., Gajjar, A., Gilbert, M. R., . . . Gilbertson, R. J. (2019). Challenges to curing primary brain tumours. *Nature Reviews Clinical Oncology, 16*, 509–520. doi: 10.1038/s41571-019-0177-5

[11] NCI (n.d.). Definition of Malignancy—NCI dictionary of cancer terms. (https://www.cancer.gov/ publications/dictionaries/cancer-terms/def/malignancy)

[12] Mckinney, P. A. (2004). Brain tumours: Incidence, survival, and aetiology. *Journal of Neurology, Neurosurgery, and Psychiatry, 75*(Suppl. II), ii12–ii17. doi: 10.1136/jnnp.2004.040741

[13] Laquintana, V., Trapani, A., Denora, N., Wang, F., Gallo, J. M., & Trapani, G. (2009). New strategies to deliver anticancer drugs to brain tumors. *Expert Opinion on Drug Delivery, 6*, 1017–1032. doi: 10.1517/17425240903167942

[14] Jovčevska, I. (2019). Genetic secrets of long-term glioblastoma survivors. *Bosnian Journal of Basic Medical Sciences, 19*(2), 116.

[15] Hanif, F., Muzaffar, K., Perveen, K., Malhi, S. M., & Simjee, S. U. (2017). Glioblastoma multiforme: A review of its epidemiology and pathogenesis through clinical presentation and treatment. *Asian Pacific Journal of Cancer Prevention, 18*, 3–9. doi: 10.22034/APJCP.2017.18.1.3

[16] Weller, M., Wick, W., Aldape, K., Brada, M., Berger, M., Pfister, S. M., . . . Reifenberger, G. (2015). Glioma. *Nature Reviews Disease Primers, 1*(1), 1–18.

[17] van Tellingen, O., Yetkin-Arik, B., de Gooijer, M. C., Wesseling, P., Wurdinger, T., & de Vries, H. E. (2015). Overcoming the blood-brain tumor barrier for effective glioblastoma treatment. *Drug Resistance Updates, 19*, 1–12. doi: 10.1016/j.drup.2015.02.002

[18] Davis, M. E. (2016). Glioblastoma: overview of disease and treatment. *Clinical Journal of Oncology Nursing, 20*, 2–8. doi: 10.1188/16.CJON.S1.2-8

[19] Birzu, C., French, P., Caccese, M., Cerretti, G., Idbaih, A., Zagonel, V., & Lombardi, G. (2021). Recurrent glioblastoma: From molecular landscape to new treatment perspectives. *Cancers, 13*, 47. doi: 10.3390/ cancers13010047

[20] Hambardzumyan, D., & Bergers, G. (2015). Glioblastoma: Defining tumor niches. *Trends in Cancer, 1*, 252–265. doi: 10.1016/j.trecan.2015.10.009

[21] Ostrom, Q. T., Cote, D. J., Ascha, M., Kruchko, C., & Barnholtz-Sloan, J. S. (2018). Adult glioma incidence and survival by race or ethnicity in the United States from 2000 to 2014. *JAMA Oncology, 4*, 1254–1262. doi: 10.1001/jamaoncol.2018.1789

[22] Louis, D. N., Perry, A., Wesseling, P., Brat, D. J., Cree, I. A., Figarella-Branger, D., . . . Ellison, D. W. (2021). The 2021 WHO classification of tumors of the central nervous system: A summary. *Neuro-Oncology, 23*(8), 1231–1251.

[23] Lamszus, K. (2004). Meningioma pathology, genetics, and biology. *Journal of Neuropathology & Experimental Neurology, 63*(4), 275–286.

[24] Löw, S., Han, C. H., & Batchelor, T. T. (2018). Primary central nervous system lymphoma. *Therapeutic Advances in Neurological Disorders* 11.

[25] Hirato, J., & Nakazato, Y. (2001). Pathology of pineal region tumors. *Journal of Neuro-Oncology, 54*(3), 239–249.

[26] Wiemels, J., Wrensch, M., & Claus, E. B. (2010). Epidemiology and etiology of meningioma. *Journal of Neuro-Oncology, 99*(3), 307–314.

[27] Gal, T. J., Shinn, J., & Huang, B. (2010). Current epidemiology and management trends in acoustic neuroma. *Otolaryngology—Head and Neck Surgery, 142*(5), 677–681.

[28] Anjum, K., Shagufta, B. I., Abbas, S. Q., Patel, S., Khan, I., Shah, S. A. A., . . . ul Hassan, S. S. (2017). Current status and future therapeutic perspectives of glioblastoma multiforme (GBM) therapy: A review. *Biomedicine & Pharmacotherapy, 92*, 681–689.

[29] Roda, E., & Bottone, M. G. (2022). Brain cancers: New perspectives and Ttherapies. *Frontiers in Neuroscience, 16*.

[30] Asa, S. L., & Ezzat, S. (2002). The pathogenesis of pituitary tumours. *Nature Reviews Cancer, 2*(11), 836–849.

[31] Wesseling, P., van den Bent, M., & Perry, A. (2015). Oligodendroglioma: Pathology, molecular mechanisms and markers. *Acta Neuropathologica, 129*(6), 809–827.

[32] Burger, P. C. (2002). What is an oligodendroglioma? *Brain Pathology*, *12*(2), 257–259.

[33] Müller, H. L. (2014). Craniopharyngioma. *Endocrine Reviews*, *35*(3), 513–543.

[34] Grant, J. W., Gallagher, P. J., & Hedinger, C. (1988). Haemangioblastoma. *Acta Neuropathologica*, *76*(1), 82–86.

[35] Reya, T., Morrison, S. J., Clarke, M. F., & Weissman, I. L. (2001). Stem cells, cancer, and cancer stem cells. *Nature*, *414*, 105–111.

[36] Quail, D. F., & Joyce, J. A. (2013). Microenvironmental regulation of tumor progression and metastasis. *Nature Medicine*, *19*(11), 1423–1437. doi: 10.1038/nm.3394

[37] Yeole, B. B. (2008). Trends in the brain cancer incidence in India. *Asian Pacific Journal of Cancer Prevention*, *9*(2), 267–270.

[38] Kleihues, P., H. Ohgaki, R. H. Eibl, M. B. Reichel, L. Mariani, M. Gehring…M. Schwab. (1994). Type and frequency of p53 mutations in tumors of the nervous system and its coverings. In: Wiestler, O. D., Schlegel, U., Schramm, J. (eds.) *Molecular Neuro-oncology and Its Impact on the Clinical Management of Brain Tumors. Recent Results in Cancer Research*, vol. 135. Springer, Berlin, Heidelberg. https://doi.org/10.1007/978-3-642-85039-4_4

[39] Hodson, T. S., Nielsen, S. M., Lesniak, M. S., Lukas, R. V. (2016). Neurological management of Von Happel-Lindau disease. *Neurologist (Review)*, *21*(5), 73–78.

[40] Hourigan, C. S. (2006). The molecular basis of coeliac disease. *Clinical and Experimental Medicine*, *6*(2), 53–59.

[41] Clapp, R. W., Jacobs, M. M., & Loechler, E. L. (2008). Environmental and occupational causes of cancer: New evidence 2005–2007. *Reviews on Environmental Health*, *23*(1), 1–38.

[42] Morgan, L. L., Miller, A. B., Sasco, A., & Davis, D. L. (2015). Mobile phone radiation causes brain tumors and should be classified as a probable human carcinogen (2A). *International Journal of Oncology*, *46*(5), 1865–1871.

[43] Soffritti, M., Belpoggi, F., Manservigi, M., Tibaldi, E., Lauriola, M., Falcioni, L., & Bua, L. (2010). Aspartame administered in feed, beginning prenatally through life span, induces cancers of the liver and lung in male Swiss mice. *American Journal of Industrial Medicine*, *53*, 1197–1206.

[44] Clarke, M. F. (2004). Neurobiology: At the root of brain cancer. *Nature*, *432*(7015), 281–282. doi: 10.1038/432281a

[45] Lessard, J., & Sauvageau, G. (2003). Bmi-1 determines the proliferative capacity of normal and leukaemic stem cells. *Nature*, *423*(6937), 255–260.

[46] van Lohuizen, M., Frasch, M., Wientjens, E., & Berns, A. (1991). Sequence similarity between the mammalian *bmi-1* proto-oncogene and the *Drosophila* regulatory genes *Psc* and *Su(z)2*. *Nature*, *353*, 353–355.

[47] *Brain tumors*. www.merckmanuals.com/home/brain,-spinal-cord,-and-nerve-disorders-of-the-nerves-system/braintumors

[48] Kahn, K., & Finkel, A. (2014). Current review of headache and brain tumor. *Current Pain and Headache Reports*, *18*(6), 421.

[49] Mood swings and cognitive changes. *American Brain Tumor Association*. Retrieved August 3, 2016, from http://www.abta.org/brain-tumorinformation/symtoms/mood-swings-html

[50] Caleb, J. *Brain tumor symptoms. Miles for hope*. Brain Tumor Foundation. Retrieved August 3, 2016, from http://web:.archive.org/web/20160814084958/; http://milesforhope.org/index.php/facts/brain.tumor.symptoms

[51] Yekula, A., Yekula, A., Muralidharan, K., Kang, K., Carter, B., & Balaj, L. (2020). Extracellular vesicles in glioblastoma tumor microenvironment. *Frontiers in Immunology*, *10*, 3137. doi: 10.3389/fimmu.2019.03137

[52] Dahan, P., Martinez Gala, J., Delmas, C., Monferran, S., & Malric, L. (2014). Ionizing radiations sustain glioblastoma cell dedifferentiation to a stem-like phenotype through survivin: Possible involvement in radioresistance. *Cell Death & Disease*, *5*, e1543. doi: 10.1038/cddis.2014.509

[53] van Tellingen, O., Yetkin-Arik, B., de Gooijer, M. C., Wesseling, P., Wurdinger, T., & de Vries, H. E. (2015). Overcoming the blood-brain tumor barrier for effective glioblastoma treatment. *Drug Resistance Updates*, *19*, 1–12. doi: 10.1016/j.drup.2015.02.002

[54] Brahm, C. G., van Linde, M. E. H, Enting, R. H., Schuur, M., Otten, R. H. J., Heymans, M. W., . . . Walenkamp, A. M. E. (2020). The current status of immune checkpoint inhibitors in neuro-oncology: A systematic review. *Cancers*, *12*, 586. doi: 10.3390/cancers12030586

[55] von Neubeck, C., Seidlitz, A., Kitzler, H. H., Beuthien-Baumann, B., & Krause, M. (2015). Glioblastoma multiforme: Emerging treatments and stratification markers beyond new drugs. *The British Journal of Radiology*, *88*, 20150354. doi: 10.1259/bjr.20150354

[56] Chen, X., Zhang, M., Gan, H., Wang, H., Lee, J. H., Fang, D., . . . Zhang, Z. (2018). A novel enhancer regulates MGMT expression and promotes temozolomide resistance in glioblastoma. *Nature Communications, 9*, 2949. doi: 10.1038/s41467-018-05373-4

[57] Lee, C. H., Cho, J., & Lee, K. (2020). Tumour regression via integrative regulation of neurological, inflammatory, and hypoxic tumour microenvironment. *Biomolecules & Therapeutics, 28*, 119–130. doi: 10.4062/biomolther.2019.135

[58] Lange, F., Hartung, J., Liebelt, C., Boisserée, J., Resch, T., Porath, K., . . . Kirschstein, T. (2020). Perampanel add-on to standard radiochemotherapy in vivo promotes neuroprotection in a rodent F98 glioma model. *Frontiers in Neuroscience, 14*, 598266.

[59] Ferrari, B., Roda, E., Priori, E. C., De Luca, F., Facoetti, A., Ravera, M., . . . Bottone, M. G. (2021). A new platinum-based prodrug candidate for chemotherapy and its synergistic effect with hadrontherapy: Novel strategy to treat glioblastoma. *Frontiers in Neuroscience, 15*, 589906.

2 Epidemiology of Brain Cancer

Xiaodong Li¹, Dipro Mukherjee², Sayantani Garai²,
Mansi Agarwal³, Sanchari Das², Dibyajit Lahiri²,
and Moupriya Nag²*

¹Department of Neurosurgery, Shengjing Hospital of China Medical
University, Shenyang, Liaoning, China
²Department of Biotechnology, University of Engineering & Management,
Kolkata, West Bengal, India
³Department of Bioscience & Bioengineering, Indian Institute of
Technology, Jodhpur
*Corresponding author

2.1 INTRODUCTION

Epidemiologic research on glioma has investigated numerous risk factors over the past few decades. Despite their rarity, brain tumors significantly contribute to morbidity, commonly affecting children and generally having a poor prognosis [1]. Glioblastoma patients, in particular, have a very poor prognosis due to the cancer's severe resistance to radiation and chemotherapy. Less than 3% of glioblastoma patients survive beyond three years, with the majority succumbing within 9 to 12 months. Genetic alterations associated with brain tissue tumors may inspire the development of novel therapeutic approaches, such as gene therapy [2].

Malignant brain tumors are rare, accounting for about 2% of all adult malignancies. In the UK, approximately 4,400 people are diagnosed with a brain tumor each year, compared to around 40,000 women diagnosed with breast cancer and about 25,000 men diagnosed with prostate cancer. On average, 7 brain tumors are diagnosed per 100,000 people annually. Nervous system tumors, which are more common in men than women, constitute about 3% of all cancers globally [3]. The incidence of brain and CNS cancer varies significantly worldwide, with the highest rates found primarily in Europe and the lowest in Asia, differing by a factor of five.

This chapter presents the descriptive epidemiology of brain tumors, including information on risk factors such as metals, hereditary susceptibility, gender, N-nitroso compounds, ionizing and non-ionizing radiation, immunological function (including the ability to fend off infections and allergies), and well-established neurocarcinogens [4].

2.2 INCIDENCE AND MORTALITY

The highest prevalence of brain tumors is observed in developed nations, which may be attributed to better registration processes that account for benign tumors. Similar to almost all other adult malignancies, the incidence of brain tumors increases with age, starting around the age of 30. In elderly patients, comorbid conditions such as strokes may mask the symptoms of brain tumors, or physicians may be reluctant to conduct rigorous examinations [5]. Men are more likely than women to be diagnosed with brain tumors, with a male-to-female ratio of 1.5. Additionally, primary malignant brain tumors are more common in men, while women are more likely to develop non-malignant tumors like meningiomas. However, meningiomas are actually more common in women than in men.

In 2020, primary CNS and brain tumors were expected to be the leading cause of death for 251,330 people worldwide. The age-standardized male mortality rate for primary malignant brain tumors is 2.8 per 100,000, while the female mortality rate is 2.0 per 100,000. Similar to incidence rates, estimated

DOI: 10.1201/9781003519706-2

mortality rates are higher in more developed nations (4.1 per 100,000 for males and 2.7 per 100,000 for females) compared to less developed nations (2.2 per 100,000 for males and 1.6 per 100,000 for females) [6]. In India, the prevalence of central nervous system tumors ranges between 5 and 10 per 100,000 people and is increasing. These tumors constitute 2% of all malignancies, with male and female mortality rates for primary malignant brain tumors at 4.6 and 2.7 per 100,000, respectively [7].

Evidence from national cancer registries indicates that the epidemiology of brain tumors differs between children and adults. For example, in Sweden, high-grade glioma (30.5%) and meningioma (29.4%) are the most prevalent types of adult primary brain tumors, whereas medulloblastoma (23.5%) and low-grade glioma (31.7%) are the most common types among children aged 15 years and under [8]. The incidence and mortality of brain tumors are increasing in industrialized nations relative to other parts of the world due to factors such as increased life expectancy, urbanization, and lifestyle changes.

2.3 RISK FACTORS

A risk factor is anything that raises a person's chances of developing a brain tumor. Although risk factors often influence the likelihood of tumor development, most do not directly cause a tumor. It is possible for individuals without any known risk factors to develop brain tumors, while others with many risk factors may never develop one [9]. Being aware of your risk factors and discussing them with your doctor can help you make more informed decisions about your health.

2.3.1 FAMILY HISTORY

An individual with a family history of brain tumors may have inherited a genetic mutation that increases their risk of developing the illness. One notable condition linked to inherited genetic mutations is neurofibromatosis, which arises from mutations in the NF1 or NF2 genes. This condition elevates the risk of developing tumors such as schwannomas, gliomas, and meningiomas. Another condition, Turcot syndrome, involves mutations in the APC, MLH1, or PMS2 genes and increases the likelihood of developing gliomas and brain tumors [10]. These findings underscore the importance of considering family history when assessing the risk of brain tumors and other neurological conditions.

2.3.2 GENDER

In general, men have a higher risk of developing brain tumors than women. Medulloblastomas are more commonly found in males, whereas meningiomas, a particular form of brain tumor, are more frequent in females. Overall, 58% of diagnosed brain and CNS cancers are found in females, compared to 42% in males. Additionally, non-malignant brain tumors are substantially more common in females (64%) than in males (36%). Conversely, males are slightly more likely than females (56%) to have malignant brain tumors [11].

Different brain tumors have varying incidence rates. For instance, men are more likely to be diagnosed with glioblastoma, whereas women are more likely to be diagnosed with meningioma. With one exception, males generally have higher fatality rates from malignant brain tumors than females [12].

2.3.3 AGE

Age significantly influences the frequency of brain cancer, with individuals aged 65 and older experiencing higher incidence rates. The impact of age varies depending on the tumor's cell type and location. For example, medulloblastomas are extremely rare in adults but more common in children, whereas gliomas are more frequently diagnosed in adults. While brain tumors can develop at any age, meningiomas and craniopharyngiomas are much more common in adults over the age of 50 [13].

Although anyone can develop a brain tumor, children and older adults are at higher risk. Advancing age is the most significant risk factor for cancer in general and for many specific cancer types. Cancer incidence rates increase consistently with age, from less than 25 cases per 100,000 in individuals under 20 to approximately 350 per 100,000 in the 45–49 age group, and over 1,000 per 100,000 in those aged 60 and older. However, cancer can be diagnosed at any age. For instance, around one-fourth of bone cancer cases occur in individuals under 20, with the highest frequency in children and adolescents. Additionally, 12% of brain and other nervous system malignancies are found in children and adolescents, compared to 1% of all cancers [14].

2.4   EXPOSURE TO CHEMICALS AND RADIATION

Chemical exposure can lead to brain damage. When hazardous substances are released into the environment, individuals may come into contact with them. These substances can originate from incinerators, factories, industrial facilities, landfills, tanks, or drums. Exposure can occur through ingestion, skin contact, or inhalation [15]. The harmful effects of chemicals on the body vary, with children being particularly vulnerable. Young people face higher risks from chemical exposure compared to adults. Pregnant women are also at greater risk, as exposure can irreversibly damage the developing organs of the fetus. Exposure to certain industrial chemicals or solvents has been linked to an increased risk of brain cancer [16]. Research indicates that individuals working in the pharmaceutical, rubber, and oil refining industries are more likely to develop specific types of brain tumors. The risk of brain cancer also increases with radiation therapy, particularly when exposure occurs early in life. Radiation, often from therapy used to treat other conditions, is the most established environmental risk factor for brain cancers. For example, low-dose radiation therapy was once commonly used to treat children with scalp ringworm, a fungal infection, before its dangers were understood [17]. It was later found that this practice increased the risk of certain brain tumors as the patients aged. Currently, most radiation-induced brain tumors are caused by head radiation used to treat other cancers, particularly in individuals who received such treatment for leukemia during childhood. Radiation can be either ionizing or non-ionizing [18].

2.5   IONIZING RADIATION EXPOSURE

It is well established that specific types and amounts of ionizing radiation induce brain tumors. Ionizing radiation generally has a stronger correlation with the risk of meningioma than with the risk of glioma [19]. Exposure at younger ages shows a stronger positive correlation with glioma risk compared to exposure at older ages. Although no differences in meningioma risk were observed according to sex, age at exposure, time since exposure, or achieved age, it is possible that the sample size was insufficient to detect these differences [20]. The role of ionizing radiation in the genesis of brain cancers varies according to the specific tumor characteristics. Therapeutic doses of ionizing radiation remain one of the few known risks for brain tumors. For instance, irradiation used to treat childhood malignancies and leukemia, as well as the now-discontinued low-dose radiation therapy for Tinea capitis and other skin conditions in children, has been associated with an elevated risk of brain tumors extending into adulthood. In utero exposure does not appear to affect the risk to the developing embryo [21]. Diagnostic X-rays do not seem to be associated with gliomas; however, some studies have found a connection between whole-mouth dental X-rays and meningiomas.

2.6   NON-IONIZING RADIATION EXPOSURE

Non-ionizing radiation and electromagnetic fields are both subsets of electromagnetic radiation, encompassing different wavelengths and energy types within the electromagnetic spectrum [22]. Non-ionizing radiation includes wavelengths ranging from 100 Hz to 300 GHz. Recent research has focused on mobile phones, as these radio frequency (RF) exposures occur close to the head and brain, despite their ubiquity [23]. Understanding the relationship between non-ionizing radiation exposure—specifically, exposures to electromagnetic fields at RFs or extremely low frequencies—and the development of primary brain tumors is crucial. The ambiguous link between cell phone use and both gliomas and meningiomas is particularly noteworthy [24].

As mobile phone use increases, the risk of brain tumors in individuals over 20 years of age also rises. Concerns associated with this exposure include headaches, sleep disruptions, short-term memory impairment, elevated blood pressure, and strokes. The relative risk was found to be 0.15 for men and 0.05 for women for every 30 seconds spent using mobile phones. Therefore, it is advisable to maintain a safe distance from mobile phones during use, keep a safe distance from antennas, and utilize all available measures to protect oneself from harmful radiation. These tumors, whether benign or malignant, can significantly damage the brain and spinal cord, and primary brain tumors may metastasize to other parts of the body [25].

2.7   N-NITROSO COMPOUNDS

Experimental carcinogens such as N-nitroso compounds (NOCs) and their precursors have been investigated as potential causes of brain cancers in humans. NOCs are known to be carcinogenic to several animal organs and can induce cancer in approximately 40 different animal species, including higher primates. These compounds may contribute to the development of various human malignancies, including those of the brain, esophagus, nasopharynx, and colon. Among the NOCs, N-nitrosodimethylamine has been classified as "possibly carcinogenic to humans" [26].

NOCs can be categorized into two main groups based on their chemical structure: nitrosamines and nitrosamides. Nitrosamines have been shown to induce cancer in animal studies, but there is no evidence linking them to brain or spinal cord malignancies in humans. The metabolic activation of nitrosamines is believed to be facilitated by a group of enzymes known as cytochromes P-450, specifically a sub-family involved in ethanol detoxification.

On the other hand, nitrosamides are direct-acting agents that can form DNA adducts and have been demonstrated to be potent nervous system carcinogens across various species. Unlike nitrosamines, nitrosamides do not require metabolic activation and can induce brain tumors in animal models when administered directly [27].

2.8   HEAD INJURY AND SEIZURES

The potential relationship between brain tumors and severe head injury has been extensively studied. While several studies suggest an association between head trauma and meningioma, no such link has been established for glioma. Seizures, which are more common in children and younger individuals, can sometimes be triggered by abnormal brain wiring that leads to electrical surges, causing brain cells to fire erratically and affecting the entire brain [28].

One of the major challenges in studying this relationship is "recall bias," where individuals with brain tumors are more likely to remember instances of head trauma compared to the general population. To mitigate this bias, some epidemiological studies have utilized medical records to investigate the connection between head trauma and the subsequent development of brain tumors. However, these studies have largely failed to demonstrate a significant relationship. The pervasive issue of recall bias complicates the analysis and interpretation of research in this area, making it difficult to draw definitive conclusions about the increased risk [29].

2.9   ALCOHOL AND SMOKING

According to the data, the likelihood of developing alcohol-related cancers increases with the amount of alcohol consumed, particularly with consistent long-term consumption. Even light drinkers and binge drinkers face a slightly elevated risk of several types of cancer. Globally, approximately 3.6% of all cancers, or 389 out of every 1,000 cases, are associated with alcohol use [30]. Furthermore, alcohol consumption is linked to poorer survival rates and increased tumor growth in cancer patients. An Australian study found that higher alcohol intake correlates with a lower chance of survival after a brain tumor diagnosis.

Smoking is one of the leading causes of human cancer [31]. In high-income countries, smoking accounts for 25–30% of all cancer deaths. The carcinogenic effect of smoking is due to the binding of nitrous compounds and polycyclic aromatic hydrocarbons to surfaces. Although many components of tobacco smoke do not cross the blood–brain barrier, smoking still significantly impacts cancer risk. Both smokers and children born to mothers who smoke have a higher risk of developing brain tumors. Smoking damages DNA and impairs the proteins responsible for DNA repair. Additionally, smoking hinders the body's metabolic detoxification processes, which neutralize and expel harmful toxins [32, 33].

2.10   VIRUSES, ALLERGIES, AND INFECTIONS

There is limited epidemiological evidence suggesting that certain viruses, such as retroviruses, papovaviruses, and adenoviruses, can induce brain tumors in experimental animal models. Initially, there was concern that live polio vaccines contaminated with SV40 might increase the risk of brain tumors; however, more comprehensive studies later refuted this hypothesis [34]. Direct examination of brain tumor tissues for viral etiology has detected viral DNA sequences in a subset of patients across various pathological series [35]. Despite these findings, the mechanisms by which viruses might initiate malignant transformations remain unclear, necessitating further research to elucidate the potential role of viruses in brain tumor development.

Atopic conditions, including allergies, asthma, and eczema, serve as indicators of immune system functioning and have been shown to be "protective" against gliomas in several independent studies from different countries. Glioma patients tend to report fewer atopic symptoms compared to controls, suggesting a potential causal role for immunologic factors in glioma development [36]. In utero infections with influenza and varicella have been suggested as risk factors for brain tumors, though the evidence is weak [37].

Recent epidemiological research on children diagnosed with brain tumors in Northwest England has revealed geographic distributions indicative of an infectious etiology for certain tumor types. The clustering of cases by time, geography, and seasonality of diagnosis suggests that infections might be risk factors [38]. The relationship between infections, immune responses, and brain tumor development is a critical area for further investigation [39].

Brain cancer can originate in the brain or its surrounding structures or metastasize from other body parts. Exposure to certain viruses, including Epstein-Barr virus (EBV), SV40, and cytomegalovirus (CMV), may increase the risk of developing brain cancer [40]. Most viruses are microscopic and consist of a small number of genes (DNA or RNA) encased in proteins. They replicate by infiltrating living cells and commandeering the cell's machinery. Some viruses achieve this by integrating their DNA (or RNA) into the host cell's DNA, potentially pushing the host cell toward cancer development [41].

Research suggests that SV40 infection might increase the risk of mesothelioma, brain tumors, and bone malignancies. Additionally, EBV can induce chromosomal instability and centrosome amplification without establishing a persistent infection, raising the possibility of malignancies developing independently of the viral genome [42].

2.11   WEAK IMMUNE SYSTEM

The immune system plays a crucial role in protecting the body from parasites, fungi, viruses, and bacteria that can cause illness. It is responsible for defending against numerous illnesses and infections. Individuals who maintain a healthy diet generally have robust immune systems [43]. Illness in an individual often indicates a compromised immune system. The immune system comprises white blood cells, antibodies, and various organs and lymph nodes.

People with compromised immune systems are more susceptible to brain and spinal cord lymphomas, known as primary CNS lymphomas [44]. Lymphomas are cancers of lymphocytes, a type of white blood cell that combats disease. Primary CNS lymphomas are less common than lymphomas that develop outside the brain. A compromised immune system can be hereditary (present at birth), result from previous cancer treatments, be due to medications taken to prevent organ rejection, or be associated with conditions such as acquired immunodeficiency syndrome [45–49].

2.12 NEUROCARCINOGENS

The most common source of neurocarcinogens for human exposure is food, particularly vegetables and cured meats. Certain alkylating chemicals, such as ethyl and methyl nitrosourea, are known transplacental carcinogens, especially in rat brain tumors [50–54]. These chemicals are potent initiators of the carcinogenic process due to their ability to cross the blood–brain barrier and their mutagenic potential. NOCs found in food and the environment, along with antioxidant consumption, have been investigated in humans as potential causes of brain tumors [55–59]. Antioxidants are sourced from fresh fruits and vegetables, supplements, and endogenous metabolic processes. However, the literature presents conflicting evidence regarding the role of these substances or other dietary elements, such as vitamin supplements, in the development of brain tumors.

Nitrate concentrations in drinking water have also been studied, but no consistent associations have been established. Aspartame, a low-calorie sweetener widely used in various food products for over 15 years, has been hypothesized to contribute to the etiology of several brain tumors, primarily based on laboratory findings [60–63]. Cadmium, a category I carcinogen, is the metal most likely to cause brain tumors. It is linked to cancers of the lung, brain, bladder, breast, liver, and stomach and is prevalent in the environment and commonly used in manufacturing consumer goods. The primary exposure factors are food, smoking, and occupation [64–69].

The relationship between drinking water and brain tumors may be influenced by environmental contaminants such as chloroethane, produced during sewage and wastewater treatment, or nitrate/nitrite contamination that leaches into drinking water supplies [70].

2.13 CONCLUSION

Adult brain tumors are rare and often carry a poorer prognosis compared to many other cancers. Reports of rising incidence trends should be interpreted with caution, as they may be influenced by changes in diagnostic and clinical procedures. The profile of tumor types differs between children and adults, with children having improved survival rates [71]. Beyond genetic predisposition, the most well-established environmental risk factor for brain tumors is exposure to high doses of ionizing radiation. However, most individuals with risk factors for brain cancer do not develop the disease. Research into the roles of immune systems and infections may yield valuable insights [72–79].

Although brain tumors are generally serious, treatment is possible. With appropriate care and therapy, some cases can be successfully managed or even cured. Early diagnosis is crucial for effective cancer treatment. Awareness of risk factors can encourage individuals to adopt preventive measures, such as quitting smoking, maintaining a healthy diet, and exercising regularly.

REFERENCES

[1] Kaderali Z, Lamberti-Pasculli M, Rutka JT. The changing epidemiology of paediatric brain tumors: A review from the Hospital for Sick Children. Childs Nerv Syst 2009;25:787–93.

[2] Siegel R, Naishadham D, Jemal A. Cancer statistics, 2013. CA Cancer J Clin 2013;63:11–30.

[3] Bauchet L, Rigau V, Mathieu-Daude H, Fabbro-Peray P, Palenzuela G, Figarella-Branger D, et al. Clinical epidemiology for childhood primary central nervous system tumors. J Neuro Oncol 2009;92:87–98.

[4] Ostrom QT, Gittleman H, Farah P, Ondracek A, Chen Y, Wolinsky Y, et al. CBTRUS statistical report: Primary brain and central nervous system tumors diagnosed in the United States in 2006–2010. Neuro Oncol 2013;15(Suppl 2):ii1–56.

[5] Stokland T, Liu JF, Ironside JW, Ellison DW, Taylor R, Robinson KJ, et al. A multivariate analysis of factors determining tumor progression in childhood low-grade glioma: A population-based cohort study (CCLG CNS9702). Neuro Oncol 2010;12:1257–68.

[6] Freeman CR, Farmer JP. Pediatric brain stem gliomas: A review. Int J Radiat Oncol Biol Phys 1998; 40:265–71.

[7] Hargrave D, Bartels U, Bouffet E. Diffuse brainstem glioma in children: Critical review of clinical trials. Lancet Oncol 2006;7:241–8.

[8] Louis DN, Wiestler OD, Cavanee WK, editors. WHO classification of tumors of the central nervous system. Lyon: International Agency for Research on Cancer; 2007.

[9] Smoll NR. Relative survival of childhood and adult medulloblastomas and primitive neuroectodermal tumors (PNETs). Cancer 2012;118:1313–22.

[10] Kleihues P, Burger PC, Scheithauer BW. The new WHO classification of brain tumors. Brain Pathol 1993;3:255–68.

[11] Rutkowski S, von Hoff K, Emser A, Zwiener I, Pietsch T, Figarella-Branger D, et al. Survival and prognostic factors of early childhood medulloblastoma: An international meta-analysis. J Clin Oncol 2010;28:4961–8.

[12] Kool M, Korshunov A, Remke M, Jones DT, Schlanstein M, Northcott PA, et al. Molecular subgroups of medulloblastoma: An international meta-analysis of transcriptome, genetic aberrations, and clinical data of WNT, SHH, Group 3, and Group 4 medulloblastomas. Acta Neuropathol 2012;123:473–84.

[13] Woehrer A, Slavc I, Waldhoer T, Heinzl H, Zielonke N, Czech T, et al. Incidence of atypical teratoid/rhabdoid tumors in children: A population-based study by the Austrian Brain Tumor Registry, 1996–2006. Cancer 2010;116:5725–32.

[14] Ostrom QT, Chen Y, Blank PD, Ondracek A, Farah P, Gittleman H, et al. The descriptive epidemiology of atypical teratoid/rhabdoid tumors in the United States, 2001–2010. Neuro Oncol 2014;16:1392–9.

[15] Lafay-Cousin L, Hawkins C, Carret AS, Johnston D, Zelcer S, Wilson B, et al. Central nervous system atypical teratoid rhabdoid tumors: The Canadian Paediatric Brain Tumor Consortium experience. Eur J Cancer 2012;48:353–9.

[16] Hilden JM, Meerbaum S, Burger P, Finlay J, Janss A, Scheithauer BW, et al. Central nervous system atypical teratoid/rhabdoid tumor: Results of therapy in children enrolled in a registry. J Clin Oncol 2004;22:2877–84.

[17] von Hoff K, Hinkes B, Dannenmann-Stern E, von Bueren AO, Warmuth-Metz M, Soerensen N, et al. Frequency, risk-factors and survival of children with atypical teratoid rhabdoid tumors (AT/RT) of the CNS diagnosed between 1988 and 2004, and registered to the German HIT database. Pediatr Blood Cancer 2011;57:978–85.

[18] Athale UH, Duckworth J, Odame I, Barr R. Childhood atypical teratoid rhabdoid tumor of the central nervous system: A meta-analysis of observational studies. J Pediatr Hematol Oncol 2009;31:651–63.

[19] Lee JY, Kim IK, Phi JH, Wang KC, Cho BK, Park SH, et al. Atypical teratoid/rhabdoid tumors: The need for more active therapeutic measures in younger patients. J Neuro Oncol 2012;107:413–9.

[20] Heck JE, Lombardi CA, Cockburn M, Meyers TJ, Wilhelm M, Ritz B. Epidemiology of rhabdoid tumors of early childhood. Pediatr Blood Cancer 2013;60:77–81.

[21] Bishop AJ, McDonald MW, Chang AL, Esiashvili N. Infant brain tumors: Incidence, survival, and the role of radiation based on Surveillance, Epidemiology, and End Results (SEER) data. Int J Radiat Oncol Biol Phys 2012;82:341–7.

[22] Stefanaki K, Alexiou GA, Stefanaki C, Prodromou N. Tumors of central and peripheral nervous system associated with inherited genetic syndromes. Pediatr Neurosurg 2012;48:271–85.

[23] Hottinger AF, Khakoo Y. Neurooncology of familial cancer syndromes. J Child Neurol 2009;24:1526–35.

[24] Bourdeaut F, Miquel C, Richer W, Grill J, Zerah M, Grison C, et al. Rubinstein-Taybi syndrome predisposing to non-WNT, non-SHH, group 3 medulloblastoma. Pediatr Blood Cancer 2014;61:383–6.

[25] Yu CL, Tucker MA, Abramson DH, Furukawa K, Seddon JM, Stovall M, et al. Cause-specific mortality in long-term survivors of retinoblastoma. J Natl Cancer Inst 2009;101:581–91.

[26] Dearlove JV, Fisher PG, Buffler PA. Family history of cancer among children with brain tumors: A critical review. J Pediatr Hematol Oncol 2008;30:8–14.

[27] Searles Nielsen S, Mueller BA, Preston-Martin S, Holly EA, Little J, Bracci PM, et al. Family cancer history and risk of brain tumors in children: Results of the SEARCH international brain tumor study. Cancer Causes Control 2008;19:641–8.

[28] Hemminki K, Kyyronen P, Vaittinen P. Parental age as a risk factor of childhood leukemia and brain cancer in offspring. Epidemiology 1999;10:271–5.

[29] Yip BH, Pawitan Y, Czene K. Parental age and risk of childhood cancers: A population-based cohort study from Sweden. Int J Epidemiol 2006;35:1495–503.

[30] Johnson KJ, Carozza SE, Chow EJ, Fox EE, Horel S, McLaughlin CC, et al. Parental age and risk of childhood cancer: A pooled analysis. Epidemiology 2009;20:475–83.

[31] Weinberg CR, Wilcox AJ, Lie RT. A log-linear approach to case-parent-triad data: Assessing effects of disease genes that act either directly or through maternal effects and that may be subject to parental imprinting. Am J Hum Genet 1998;62:969–78.

[32] Lupo PJ, Nousome D, Okcu MF, Chintagumpala M, Scheurer ME. Maternal variation in EPHX1, a xenobiotic metabolism gene, is associated with childhood medulloblastoma: An exploratory case-parent triad study. Pediatr Hematol Oncol 2012;29:679–85.

[33] Chen C, Xu T, Chen J, Zhou J, Yan Y, Lu Y, et al. Allergy and risk of glioma: A meta-analysis. Eur J Neurol 2011;18:387–95.

[34] Bayley KB, Belnap T, Savitz L, Masica AL, Shah N, Fleming NS. Challenges in using electronic health record data for CER: Experience of 4 learning organizations and solutions applied. Med Care 2013;51:S80–6.

[35] Harding NJ, Birch JM, Hepworth SJ, McKinney PA. Atopic dysfunction and risk of central nervous system tumors in children. Eur J Cancer 2008;44:92–9.

[36] Roncarolo F, Infante-Rivard C. Asthma and risk of brain cancer in children. Cancer Causes Control 2012;23:617–23.

[37] Shu X, Prochazka M, Lannering B, Schuz J, Roosli M, Tynes T, et al. Atopic conditions and brain tumor risk in children and adolescents–an international case-control study (CEFALO). Ann Oncol 2014;25:902–8.

[38] Fear NT, Roman E, Ansell P, Bull D. Malignant neoplasms of the brain during childhood: The role of prenatal and neonatal factors (United Kingdom). Cancer Causes Control 2001;12:443–9.

[39] Nyari TA, Dickinson HO, Parker L. Childhood cancer in relation to infections in the community during pregnancy and around the time of birth. Int J Cancer 2003;104:772–7.

[40] Dickinson HO, Nyari TA, Parker L. Childhood solid tumors in relation to infections in the community in Cumbria during pregnancy and around the time of birth. Br J Cancer 2002;87:746–50.

[41] Castro F, Bermejo JL, Hemminki K. Association between number of siblings and nervous system tumors suggests an infectious etiology. Neurology 2006;67:1979–83.

[42] Shaw AK, Li P, Infante-Rivard C. Early infection and risk of childhood brain tumors (Canada). Cancer Causes Control 2006;17:1267–74.

[43] Harding NJ, Birch JM, Hepworth SJ, McKinney PA. Infectious exposure in the first year of life and risk of central nervous system tumors in children: Analysis of day care, social contact, and overcrowding. Cancer Causes Control 2009;20:129–36.

[44] Harding NJ, Birch JM, Hepworth SJ, McKinney PA. Breastfeeding and risk of childhood CNS tumors. Br J Cancer 2007;96:815–7.

[45] Andersen TV, Schmidt LS, Poulsen AH, Feychting M, Roosli M, Tynes T, et al. Patterns of exposure to infectious diseases and social contacts in early life and risk of brain tumors in children and adolescents: An International Case-Control Study (CEFALO). Br J Cancer 2013;108:2346–53.

[46] Kobayashi N, Furukawa T, Takatsu T. Congenital anomalies in children with malignancy. Paediatr Univ Tokyo 1968;16:31–7.

[47] Miller RW. Relation between cancer and congenital defects in man. N Engl J Med 1966;275:87–93.

[48] Miller RW. Childhood cancer and congenital defects. A study of U.S. death certificates during the period 1960–1966. Pediatr Res 1969;3:389–97.

[49] Agha MM, Williams JI, Marrett L, To T, Zipursky A, Dodds L. Congenital abnormalities and childhood cancer. Cancer 2005;103:1939–48.

[50] Bjorge T, Cnattingius S, Lie RT, Tretli S, Engeland A. Cancer risk in children with birth defects and in their families: A population based cohort study of 5.2 million children from Norway and Sweden. Cancer Epidemiol Biomarkers Prev 2008;17:500–6.

[51] Fisher PG, Reynolds P, Von Behren J, Carmichael SL, Rasmussen SA, Shaw GM. Cancer in children with nonchromosomal birth defects. J Pediatr 2012;160:978–83.

[52] Partap S, MacLean J, Von Behren J, Reynolds P, Fisher PG. Birth anomalies and obstetric history as risks for childhood tumors of the central nervous system. Pediatrics 2011;128:e652–7.

[53] Bjorge T, Sorensen HT, Grotmol T, Engeland A, Stephansson O, Gissler M, et al. Fetal growth and childhood cancer: A population-based study. Pediatrics 2013;132:e1265–75.

[54] Schmidt LS, Schüz J, Lähteenmäki P, Träger C, Stokland T, Gustafson G, et al. Fetal growth, preterm birth, neonatal stress and risk for CNS tumors in children: A Nordic population- and register-based case-control study. Cancer Epidemiol Biomarkers Prev 2010;19:1042–52.

[55] Milne E, Laurvick CL, Blair E, de Klerk N, Charles AK, Bower C. Fetal growth and the risk of childhood CNS tumors and lymphomas in Western Australia. Int J Cancer 2008;123:436–43.

[56] MacLean J, Partap S, Reynolds P, Von Behren J, Fisher PG. Birth weight and order as risk factors for childhood central nervous system tumors. J Pediatr 2010;157:450–5.

[57] Harder T, Plagemann A, Harder A. Birth weight and subsequent risk of childhood primary brain tumors: A meta-analysis. Am J Epidemiol 2008;168:366–73.

[58] Kleinerman RA. Cancer risks following diagnostic and therapeutic radiation exposure in children. Pediatr Radiol 2006;36(Suppl 2):121–5.

[59] Ohgaki H, Kleihues P. Epidemiology and etiology of gliomas. Acta Neuropathol 2005;109:93–108.

[60] Streffer C, Shore R, Konermann G, Meadows A, Uma Devi P, Preston Withers J, et al. Biological effects after prenatal irradiation (embryo and fetus). A report of the International Commission on Radiological Protection. Ann ICRP 2003;33:5–206.

[61] Neglia JP, Robison LL, Stovall M, Liu Y, Packer RJ, Hammond S, et al. New primary neoplasms of the central nervous system in survivors of childhood cancer: A report from the Childhood Cancer Survivor Study. J Natl Cancer Inst 2006;98:1528–37.

[62] Mellemkjaer L, Hasle H, Gridley G, Johansen C, Kjaer SK, Frederiksen K, et al. Risk of cancer in children with the diagnosis immaturity at birth. Paediatr Perinat Epidemiol 2006;20:231–7.

[63] Stalberg K, Haglund B, Axelsson O, Cnattingius S, Pfeifer S, Kieler H. Prenatal X-ray exposure and childhood brain tumors: A population-based case-control study on tumor subtypes. Br J Cancer 2007;97:1583–7.

[64] Khan S, Evans AA, Rorke-Adams L, Orjuela MA, Shiminski-Maher T, Bunin GR. Head injury, diagnostic X-rays, and risk of medulloblastoma and primitive neuroectodermal tumor: A Children's Oncology Group study. Cancer Causes Control 2010;21:1017–23.

[65] Rajaraman P, Simpson J, Neta G, Berrington de Gonzalez A, Ansell P, Linet MS, et al. Early life exposure to diagnostic radiation and ultrasound scans and risk of childhood cancer: Case-control study. BMJ 2011;342:d472.

[66] Pearce MS, Salotti JA, Little MP, McHugh K, Lee C, Kim KP, et al. Radiation exposure from CT scans in childhood and subsequent risk of leukaemia and brain tumors: A retrospective cohort study. Lancet 2012;380:499–505.

[67] Mathews JD, Forsythe AV, Brady Z, Butler MW, Goergen SK, Byrnes GB, et al. Cancer risk in 680,000 people exposed to computed tomography scans in childhood or adolescence: Data linkage study of 11 million Australians. BMJ 2013;346:f2360.

[68] Boice JD Jr, Mumma MT, Blot WJ, Heath CW Jr. Childhood cancer mortality in relation to the St Lucie nuclear power station. J Radiol Prot 2005;25:229–40.

[69] International Agency for Cancer Research. Agents classified by the IARC Monographs, Volumes 1–109. Lyon: IARC; 2014 [cited 2014 May 7]. Available from: http://monographs.iarc.fr/ENG/Classification/

[70] Ha M, Im H, Lee M, Kim HJ, Kim BC, Gimm YM, et al. Radio-frequency radiation exposure from AM radio transmitters and childhood leukemia and brain cancer. Am J Epidemiol 2007;166:270–9.

[71] Mezei G, Gadallah M, Kheifets L. Residential magnetic field exposure and childhood brain cancer: A meta-analysis. Epidemiology 2008;19:424–30.

[72] Kheifets L, Ahlbom A, Crespi CM, Feychting M, Johansen C, Monroe J, et al. A pooled analysis of extremely low-frequency magnetic fields and childhood brain tumors. Am J Epidemiol 2010;172:752–61.

[73] Elliott P, Toledano MB, Bennett J, Beale L, de Hoogh K, Best N, et al. Mobile phone base stations and early childhood cancers: Case-control study. BMJ 2010;340:c3077.

[74] Aydin D, Feychting M, Schuz J, Tynes T, Andersen TV, Schmidt LS, et al. Mobile phone use and brain tumors in children and adolescents: A multicenter case-control study. J Natl Cancer Inst 2011;103:1264.

[75] McKean-Cowdin R, Pogoda JM, Lijinsky W, Holly EA, Mueller BA, Preston-Martin S. Maternal prenatal exposure to nitrosatable drugs and childhood brain tumors. Int J Epidemiol 2003;32:211–7.

[76] Carozza SE, Olshan AF, Faustman EM, Gula MJ, Kolonel LN, Austin DF, et al. Maternal exposure to N-nitrosatable drugs as a risk factor for childhood brain tumors. Int J Epidemiol 1995;24:308–12.

[77] Cardy AH, Little J, McKean-Cowdin R, Lijinsky W, Choi NW, Cordier S, et al. Maternal medication use and the risk of brain tumors in the offspring: the SEARCH international case-control study. Int J Cancer 2006;118:1302–8.

[78] Schuz J, Weihkopf T, Kaatsch P. Medication use during pregnancy and the risk of childhood cancer in the offspring. Eur J Pediatr 2007;166:433–41.

[79] Chuang CH, Doyle P, Wang JD, Chang PJ, Lai JN, Chen PC. Herbal medicines during pregnancy and childhood cancers: An analysis of data from a pregnancy cohort study. Pharmacoepidemiol Drug Saf 2009;18:1119–20.

3 Neuro-Ophthalmology of Brain Tumors

Xiaodong Li[1], Dipro Mukherjee[2], Sayantani Garai[2],
Mansi Agarwal[3], Sanchari Das[2], Dibyajit Lahiri[2],
and Moupriya Nag[2]*

[1]Department of Neurosurgery, Shengjing Hospital of China Medical University, Shenyang, Liaoning, China
[2]Department of Biotechnology, University of Engineering & Management, Kolkata, West Bengal, India
[3]Department of Bioscience & Bioengineering, Indian Institute of Technology, Jodhpur
*Corresponding author

3.1 INTRODUCTION

Neuro-ophthalmologists specialize in vision issues related to the nervous system, including blindness caused by damage to the brain or optic nerves transmitting visual signals from the eyes to the brain. Such damage can result from infections, toxins, tumors, strokes, or injuries [1]. Patients with difficulties controlling eye movement, leading to double vision due to misalignment or trouble seeing in certain directions, also seek treatment from neuro-ophthalmologists. Prism lenses or, in rare cases, surgery may correct this misalignment issue, known as strabismus [2].

Neuro-ophthalmologists examine patients who experience a decrease in visual acuity, visual field, or color vision due to brain or optic nerve disorders. Patients with suspected elevated intracranial pressure, which can cause vision loss and optic nerve swelling, are evaluated. Those experiencing double vision or eye movement difficulties may have conditions like myasthenia gravis or damage to brain regions, nerves, or muscles controlling eye movements [3–5]. Patients with pituitary gland or other tumors compressing visual pathways are referred to neuro-ophthalmologists before and after tumor removal to prevent vision loss, even if they are unaware of visual issues. Immediate evaluation is necessary for patients with sudden changes in pupil size, as it may indicate underlying problems. Patients with uncontrollable eyelid shaking (nystagmus) should also be assessed [3–5].

This chapter provides an overview of brain tumor classification and symptoms, focusing on types relevant to neuro-ophthalmology such as craniopharyngiomas, low-grade astrocytomas, high-grade gliomas, medulloblastoma, meningiomas, and pineoblastoma. It outlines relevant anatomy and highlights key elements of neuro-ophthalmic history and examination that are necessary for localizing lesions and understanding the clinical features that result from them [6]. Given that patients with these tumors often first consult ophthalmologists, recognizing neuro-ophthalmic symptoms can aid in tumor diagnosis.

The diagnosis and treatment of brain tumors heavily rely on neuro-ophthalmic evaluation. The afferent visual pathway consists of the retinal optic nerve, chiasm, optic tract, lateral geniculate body, optic radiations, and occipital lobe [7]. Tumors that compress this pathway often present with visual complaints. Conversely, aberrant eye movements, such as nystagmus, gaze palsies, and motility issues, involve the efferent visual pathway. Evaluation of the efferent system includes assessing the external orbit for proptosis, eyelid anomalies, and ocular movement abnormalities. By analyzing the pattern of motility disruption and visual function impairment, the tumor's location can be determined [8–10].

DOI: 10.1201/9781003519706-3

3.2   CLASSIFICATION AND SIGNIFICANCE OF BRAIN TUMOR

An abnormal mass of tissue, known as an intracranial tumor or brain tumor, is characterized by uncontrolled cell growth and reproduction that appears unaffected by the mechanisms limiting normal cell growth. While there are over 150 recognized forms of brain tumors, they are broadly categorized into primary and metastatic tumors [11].

Primary brain tumors originate in the tissues of the brain or its immediate surroundings. These tumors can be benign or malignant and can be categorized as glial (composed of glial cells) or nonglial (developed on or in brain structures, including nerves, blood vessels, and glands) [12]. The most common primary brain tumors include gliomas, meningiomas, pituitary adenomas, and nerve sheath tumors.

Metastatic brain tumors originate in other parts of the body, such as the breast or lungs, and spread to the brain typically through circulation. Malignant tumors with metastases are considered cancerous [13].

The spectrum of brain cancers encompasses aggressive tumors like glioblastoma multiforme and slower-developing tumors like meningioma. These tumors vary in their benign or malignant nature and can be primary or secondary brain tumors [14].

Symptoms of brain cancer depend on the area of the brain affected and the functional system it impacts (e.g., motor, sensory, language). For example, a tumor near the optic nerve may cause vision issues, while a tumor in the frontal lobe may affect focus and clarity of thought [15]. Weakness, numbness, or difficulty speaking may result from a tumor affecting the region responsible for motor function. Larger tumors may produce a range of symptoms due to the pressure they exert on the body.

Because the signs and symptoms of brain cancer can vary based on factors such as tumor type, size, location, patient age, and medical history, they often resemble those of other illnesses. Therefore, it's essential to seek professional medical advice for an accurate diagnosis [16].

Some early signs of brain cancer include a persistent headache that worsens over time, varies with the time of day and head position, seizures, and numbness. Other common signs may include nausea or vomiting, memory loss, muscle weakness, speech difficulties, mood or personality changes, unexplained fatigue, changes in menstrual cycle, impotence or infertility, overproduction or underproduction of breast milk, Cushing's syndrome (characterized by weight gain), high blood pressure, diabetes, and bruising [17–20].

Some individuals may experience cognitive difficulties or issues with vision, speech, or coordination. These symptoms may start mild or progress gradually. Specific symptoms associated with various types of brain tumors can vary depending on the tumor's location in the brain, such as headaches that are worse in the morning or disrupt sleep, seizures, difficulty articulating or expressing thoughts, behavioral changes, loss of balance or dizziness, weakness or paralysis in one side or region of the body, changes in vision or hearing, facial tingling or numbness, nausea, vomiting, difficulty swallowing, disorientation, and confusion [21–25].

3.3   BRAIN CANCER SPECIFIC TO NEURO-OPHTHALMOLOGY

Most symptoms and signs resulting from tumors in the cavernous sinus and orbital apex/superior orbital fissure are associated with dysfunction of the ocular motor nerves. However, trigeminal neuropathy, oculosympathetic dysfunction, and optic nerve dysfunction may also occur. The term "orbital apex syndrome" is often used when multiple cranial nerve palsies affecting eye movements and dysfunction of the optic nerve coexist [26–30]. Neuro-ophthalmic diseases can result from brain tumors, multiple sclerosis, optic nerve disorders, central nervous system conditions, traumatic brain injuries, and strokes. Common neuro-ophthalmic conditions include optic neuritis (inflammation of the optic nerve), ocular myasthenia gravis (an autoimmune condition causing muscle weakness in the eyes), and papilledema (optic nerve swelling due to intracranial pressure) [31–37].

Craniopharyngiomas are a rare type of benign (non-cancerous) brain tumor that typically originates in the pituitary gland, which regulates hormone secretion and various bodily functions. These tumors, while slow growing, can affect the pituitary gland and surrounding brain structures' functionality. Craniopharyngiomas most commonly affect young children and older adults, presenting symptoms such as headaches, fatigue, frequent urination, and gradual vision changes. Children with craniopharyngiomas may experience slower growth and development than normal [38].

Diagnosis of craniopharyngioma involves various tests and techniques to confirm its presence. A thorough examination of the patient's medical history and neurological evaluation is conducted. Blood tests may detect hormone level variations indicative of pituitary gland involvement. Imaging exams such as X-rays, magnetic resonance imaging (MRI), and computer tomography (CT) scans are performed to visualize brain structures and identify the tumor [39–44].

Incidence: Adamantinomatous and papillary craniopharyngiomas, two forms of the condition, are diagnosed in approximately three individuals per 1 million each year. They typically manifest in two age groups: children aged 5 to 15 years and adults aged 50 to 75 years. While adamantinomatous forms can occur across all age groups, papillary subtypes are predominantly found in adults [45].

Seriousness: Craniopharyngiomas constitute a significant medical condition that may require lifelong medical care. Even after surgical removal, approximately half of all tumors recur over time. Additionally, numerous medical issues associated with craniopharyngiomas persist despite tumor removal [46].

Treatment: Craniopharyngioma treatment options include:

Surgery is often recommended for individuals with craniopharyngiomas to remove the entire tumor or as much of it as possible. The type of surgery performed depends on the location and size of the tumor.

During open craniopharyngioma surgery (craniotomy), the skull is opened to access the tumor. Alternatively, transsphenoidal surgery involves inserting special surgical instruments through the nose to reach the tumor via a natural passageway without damaging the brain [47–50].

Typically, the entire tumor is removed during surgery. However, in some cases where sensitive or significant structures are nearby, surgeons may not opt to remove the entire tumor to preserve the patient's quality of life. Additional treatments may be administered after surgery in certain cases [51–55].

Radiation Therapy: Following surgery, external beam radiation therapy is often employed to treat craniopharyngiomas. This procedure utilizes high-energy beams such as protons and X-rays to eradicate tumor cells. Specialized external beam radiation technologies like intensity-modulated radiation therapy (IMRT) and proton beam therapy enable doctors to precisely shape and target the radiation beam, effectively treating tumor cells while minimizing damage to surrounding healthy tissue. In rare cases where the tumor does not affect the bundle of nerve fibers responsible for transmitting vision information from the eye to the brain, stereotactic radiosurgery, a form of radiation therapy, may be recommended [56–59]. Stereotactic radiosurgery, while technically not a surgical procedure, focuses multiple radiation beams on specific locations to destroy tumor cells. Brachytherapy, another form of radiation treatment, involves the injection of radioactive material directly into the tumor, allowing it to emit radiation from within.

Chemotherapy: Chemotherapy involves the use of drugs to kill tumor cells. To ensure that chemotherapy specifically targets the desired cells without harming nearby healthy tissue, it can be administered directly into the tumor [60].

Targeted Therapy: Targeted therapy has shown effectiveness in treating papillary craniopharyngioma, a rare subtype of craniopharyngioma. This therapy involves medications that specifically target particular defects in tumor cells, which enable their survival. In papillary craniopharyngioma, almost all cells contain a mutated gene called BRAF. Targeted therapy can potentially address this mutation. Specialized laboratory testing can determine whether the cells of papillary craniopharyngioma harbor the BRAF gene mutation or not [61–66].

Clinical Trials: Clinical trials involve the investigation of new treatments or innovative applications of existing therapies. By participating in a clinical trial, patients may gain access to the latest medications, although potential side effects may not be fully understood [67].

3.4 LOW-GRADE ASTROCYTOMAS

Compared to their higher-grade counterparts, low-grade astrocytomas are indeed rare tumors, with an estimated 1,600 cases developing in North America each year. Current studies suggest that environmental factors do not seem to play a significant role in their development, although their exact origin remains unknown. However, individuals with neurofibromatosis, a hereditary inherited illness, have a higher risk of developing these tumors [68–71].

Low-Grade Astrocytomas Encompass Three Main Tumor Forms: Pilocytic astrocytomas, pleomorphic xanthoastrocytomas, and diffuse astrocytomas. Among these, the majority are diffuse astrocytomas. Pilocytic astrocytomas, often found in youngsters, typically originate in the cerebellum, responsible for balance. One characteristic feature of pilocytic astrocytomas is their clearly defined boundaries, which can make them more treatable, especially if they are located in operable areas of the brain.

Occurrence: Low-grade astrocytomas are commonly found along the outer curve of the brain, often located at the top. Occasionally, they may also form at the base of the brain. While less frequent, astrocytomas can also be discovered in the brainstem or spinal cord [72–75].

As for their causes, the exact origin of most astrocytomas remains unknown. Therapeutic radiation is known to potentially contribute to the development of astrocytomas. While other environmental exposures have been suspected, their direct association with astrocytoma formation has not been definitively proven. Some studies suggest a hereditary component in certain cases [76].

Symptoms: The symptoms of astrocytomas can vary depending on their size and location within the brain. Common signs may include headaches, nausea and vomiting, memory loss, seizures, changes in mental status, fatigue visual problems, and other cognitive and motor impairments [77–80].

Diagnosis: The first, most common, diagnostic imaging method of choice is MRI [81]. More sophisticated imaging methods might be necessary for a tumor close to crucial brain areas (such as speech or movement control) [82]. In nearly all situations, a true tissue biopsy is necessary for a conclusive diagnosis [83]. Open surgery cannot be used to remove the tumor, but precisely guided (stereotactic) needle biopsies can. With the aid of image fusion, UCLA neurosurgeons may combine all diagnostic and imaging data in the operating room. The knowledge can then be applied during the procedure to help ensure a safer and more successful surgery.

Treatment: Efforts are made to achieve complete removal of the tumor if it is accessible through surgery. In low-grade astrocytomas, a near-complete excision (with less than 10 cc of remaining tumor) helps prevent longer-lasting tumor recurrence. However, achieving nearly full excision can be challenging for large tumors, those situated deep in the brain, or those near speech and motor centers. Surgical planning may involve the use of functional MRI (fMRI), diffusion tensor imaging for white matter tracking, and MRI spectroscopy to map critical speech and motor brain regions near the tumor. Intra-operative MRI is often utilized to aid in the most comprehensive and secure removal of low-grade astrocytomas, ensuring optimal surgical outcomes [84].

3.5 GLIOMAS OF HIGH MALIGNANCY

High-grade gliomas, pronounced "glee-OH-mas," are rapidly progressing cancers that pose a significant threat to the child's brain or spinal cord. These tumors develop when glial cells proliferate uncontrollably in the central nervous system, where they normally support neurons or nerves. Unfortunately, some high-grade gliomas cannot be cured [85]. Because they spread quickly to the brain or spinal cord, they are challenging to treat. Malignant cells can invade healthy tissue, making it difficult to remove the tumor without causing harm to surrounding healthy cells. This intertwining of cancerous and healthy cells makes gliomas particularly challenging to surgically remove without causing collateral damage to healthy tissue [86].

Occurrence: The majority of high-grade gliomas develop in either the supratentorial compartment of the brain or the brainstem. The location of the tumor is pivotal, as it directly impacts the available treatment options for the patient. For instance, surgical removal of malignancies located in the brainstem is often deemed unsafe [87].

Supratentorial Compartment: High-grade gliomas can arise from the glial cells located in the region of the brain that encompasses the cerebrum, thalamus, and hypothalamus.

Brainstem: High-grade gliomas can arise in the pons region of the brainstem, where they are referred to as diffuse intrinsic pontine gliomas [88].

Incidence: High-grade gliomas can occur in infants, children, teenagers, and adults. However, there are notable differences between high-grade gliomas in newborns and young children compared to those in adults and teenagers. These distinctions necessitate different treatment approaches for newborns and young children.

Symptoms: Depending on the location of the tumor in the child's brain or spinal cord, their age, and the rate of tumor growth, various symptoms may manifest. As the tumor enlarges and exerts pressure, symptoms become apparent [89]. Headaches are the most common symptom of high-grade glioma, especially when accompanied by nausea or vomiting. These headaches often worsen in the morning, sometimes waking the child from sleep. Other symptoms may include vision, hearing, or speech problems, difficulty with muscle coordination or balance (ataxia), confusion or memory loss, sleepiness, and lack of energy [90].

Causes: Previous radiation therapy is a known factor that may elevate the risk of high-grade glioma development in adults. While children may encounter similar external or environmental risk factors, further research is necessary to confirm this correlation. Additionally, children with certain rare inherited genetic disorders may have an increased susceptibility to high-grade gliomas. These disorders stem from abnormalities or defects in genes, which govern cell functions. However, it's noteworthy that the majority of children diagnosed with high-grade gliomas do not have a family history of the condition [91].

Diagnosis: A brain scan or imaging test can help determine if a tumor is the underlying cause of your symptoms. Imaging techniques such as MRI or CT scans enable doctors to identify tumors in the brain or spinal cord. While CT scans provide less detailed images compared to MRI, especially those enhanced with contrast, an MRI uses magnets and radio waves to produce detailed images of internal structures. During a contrast-enhanced MRI, a harmless dye is injected into the patient's arm to enhance the visibility of tumors on the image.

If a tumor is detected, a biopsy may be performed to determine if it is benign or malignant. Using a stereotactic biopsy, doctors can precisely locate the tumor and obtain a small tissue sample for analysis. Pathologists then examine the cells to identify high-grade gliomas and assess characteristics that may guide treatment decisions. For instance, identifying specific cancer cell types can inform the selection of targeted therapies tailored to the patient's condition [92–95].

Treatment: Treating the tumor can involve various methods, including surgical removal, halting its growth, and alleviating symptoms. The patient's care team typically comprises several specialists, such as a neurosurgeon, neuro-oncologist, radiation oncologist, endocrinologist, nurse practitioner, and social worker, tailored to the tumor type, its location, and other factors. Treatment options may vary and can include the expertise of these professionals:

Surgery: For high-grade gliomas that are amenable to surgical removal, surgery is the primary treatment approach. The treating physician will aim to excise as much of the tumor as possible while preserving surrounding healthy tissue. In cases involving children, focal radiation therapy is often utilized as a standard treatment. However, in consideration of promoting healthy brain development in young patients, radiation therapy is typically avoided. Instead, chemotherapy becomes the preferred option to inhibit tumor growth. Chemotherapeutic drugs are employed to target and eliminate cancer cells. While certain chemotherapy agents, such as temozolomide, show efficacy in treating specific types of adult gliomas, they may not be as effective in pediatric gliomas.

Palliative Care: Palliative care specialists play a crucial role in prioritizing the patient's quality of life throughout treatment [96–98]. They offer guidance on treatment choices that focus on enhancing the patient's well-being. Additionally, these specialists provide support to help patients cope with their diagnosis, offering options for symptom relief and overall comfort.

3.6 MEDULLOBLASTOMA

A brain tumor associated with the cerebellum is referred to as a medulloblastoma. The cerebellum governs coordinated movements and balance. Positioned at the back of the brain, in an area known as the posterior fossa, the cerebellum is situated close to the brainstem. Due to its rapid expansion, a medulloblastoma can potentially spread to other areas of the brain and spinal cord through cerebrospinal fluid (CSF).

Occurrence: Medulloblastomas typically originate in the cerebellum, which is the lowermost part of the brain located at the back of the skull (Figure 3.1). These tumors are often termed embryonal neuroepithelial tumors because they arise from fetal cells that persist after birth.

Incidence: Medulloblastoma stands as the most common malignant brain tumor among children, contributing to approximately 20% of all pediatric brain tumors [99]. While it can occur in newborns and adults, medulloblastoma predominantly impacts children. Diagnosis is most commonly made between the ages of 5 and 9. Moreover, boys exhibit a significantly higher susceptibility to developing medulloblastoma compared to girls.

Symptoms: The symptoms of medulloblastoma are influenced by the tumor's location. In cases where the tumor occurs in the cerebellum, individuals may experience challenges related to balance, walking, or fine motor skills. If the tumor obstructs the flow of CSF, it can lead to increased pressure inside the skull, a condition known as hydrocephalus [100]. Symptoms of hydrocephalus may include headaches, nausea, vomiting, blurred or double vision, extreme drowsiness, confusion, seizures, and even loss of consciousness.

Treatment: After surgical removal of the tumor in patients with medulloblastoma, radiation and chemotherapy are commonly administered to improve survival rates and minimize treatment-related side effects. Depending on the patient's prognosis, a less intensive treatment approach may be advised. However, individuals with high-risk conditions may require more extensive therapy to enhance their chances of survival.

Surgery is typically the initial step in medulloblastoma treatment [101], aiming for complete tumor removal while considering its location and the extent of safe resection.

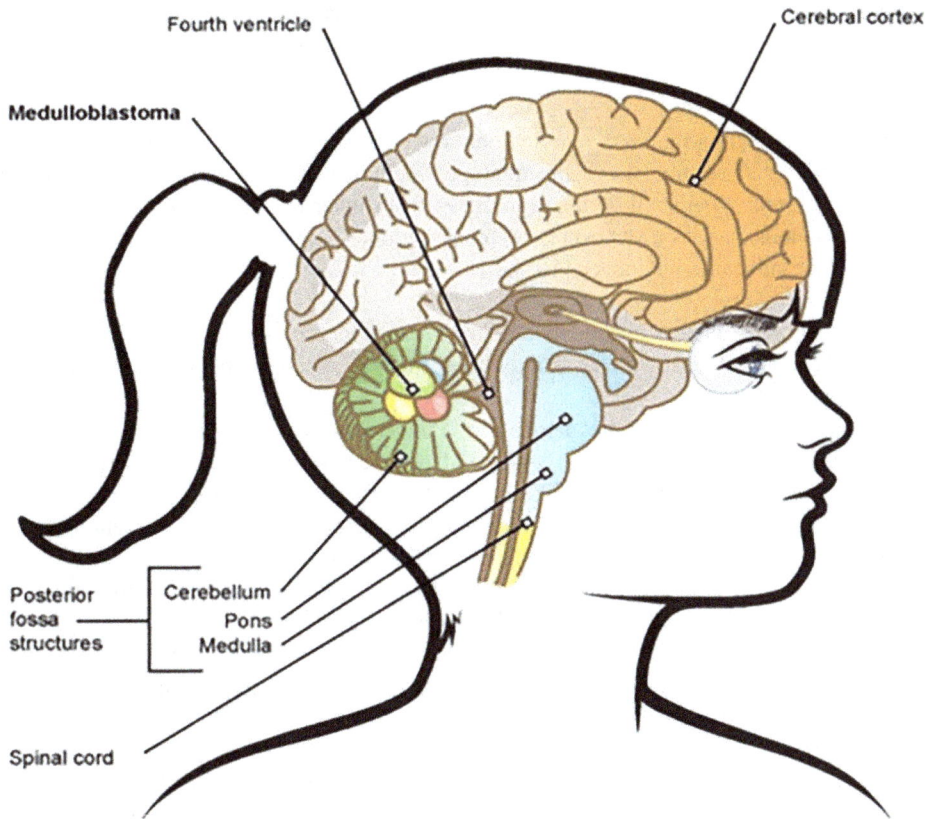

FIGURE 3.1 Anatomical illustration of the brain highlighting the location of a medulloblastoma. The diagram shows the fourth ventricle, cerebral cortex, cerebellum, pons, medulla, spinal cord, and posterior fossa structures. Medulloblastoma is indicated in the cerebellum region, illustrating its typical location within the posterior fossa. (Image courtesy St. Jude Children's Research Hospital)

Radiation therapy utilizes high-energy X-rays or alternative radiation forms like proton beam radiation to either kill cancer cells or impede their growth. The dosage is tailored based on the disease stage and risk level. Combining radiation and chemotherapy may be recommended to improve outcomes [102].

Chemotherapy employs potent medications to eradicate cancer cells or hinder their proliferation. In some cases, chemotherapy alone may effectively treat certain types of medulloblastoma. Additionally, for younger patients, chemotherapy may delay radiation treatment until they are older.

3.7   MENINGIOMAS

Meningiomas are the most common benign intracranial tumors, originating from arachnoid cap cells within the arachnoid membrane, which surrounds the brain and spinal cord resembling a spider web [103]. The meninges, comprising three protective layers covering the brain and spinal cord, include the arachnoid. Although most meningiomas are benign, delayed detection can result in slow but significant growth, posing serious life-threatening and debilitating risks. Some meningioma variants may exhibit more aggressive behavior. While many individuals develop only one meningioma, some may experience multiple tumors concurrently in different brain or spinal cord regions.

Symptoms: Symptoms of meningioma often develop gradually and may initially be subtle [104]. They can vary depending on the tumor's location in the brain or, rarely, the spine.

These symptoms may include changes in vision, such as double vision or blurred vision, ringing in the ears or hearing loss, memory loss, loss of sense of smell, seizures, weakness in the arms or legs, and difficulties with language.

The exact cause of meningiomas remains unknown. However, doctors believe that a mutation causing abnormal proliferation of certain meningeal cells leads to the formation of meningioma tumors. It is uncertain whether this mutation arises from inherited genes, hormonal factors, rare instances of previous radiation exposure, or other factors. Claims linking smartphone usage to meningiomas lack substantial evidence [105].

Risk Factors: Several factors may contribute to the development of meningiomas:

Radiation Treatment: Meningioma risk may be elevated by radiation therapy, especially when directed at the head.

Female Hormones: The higher prevalence of meningiomas in women leads medical professionals to speculate about the involvement of female hormones. Some research suggests that hormones may influence the likelihood of developing meningioma, similar to their role in breast cancer. Certain studies indicate that using oral contraceptives and hormone replacement therapy may increase the risk of meningioma.

Inherited Nervous System Disorder: Individuals with the rare disease neurofibromatosis are more likely to develop meningiomas and other brain cancers.

Obesity: Research suggests that individuals with a high body mass index may have a greater prevalence of meningiomas. However, the precise relationship between obesity and meningiomas requires further investigation [106].

Diagnosis of meningioma can be challenging due to various factors. These slow-growing tumors primarily affect adults, and their symptoms may mimic typical signs of aging, leading to misinterpretation. Additionally, symptoms associated with meningiomas can overlap with those of other medical conditions, further complicating diagnosis. It often takes several years to establish a correct diagnosis, and misdiagnosis is not uncommon.

Imaging techniques such as CT or computed axial tomography scan and MRI are typically used to identify meningiomas. In some cases, a biopsy may be necessary to confirm the diagnosis. A neurosurgeon performs the biopsy to obtain tissue samples for examination by a neuropathologist, which helps determine the tumor's grade, benign or malignant nature, and guides appropriate clinical management strategies [107, 108].

Treatment: The treatment for meningiomas includes the following:

Surgery: Surgery stands as the primary treatment option for meningiomas, offering the best chance for a cure. Typically, benign tumors with well-defined borders, meningiomas allow for complete surgical excision. The goal of surgery is to entirely remove the meningioma, including the fibers connecting it to the brain coverings and bone. However, complete excision may pose significant risks, especially if the tumor has invaded nearby veins or brain tissue. While tumor removal is paramount, preserving or improving the patient's neurological function takes precedence. To ensure safety during surgery, patients often undergo tumor embolization beforehand. Unlike cerebral angiography, which visualizes blood vessels, embolization involves the surgeon injecting a substance into the blood arteries supplying the tumor to block its blood supply. This process helps safeguard against potential complications during the surgical procedure [109].

Radiation Therapy: Radiation therapy utilizes high-energy X-rays to reduce tumor size and eliminate cancerous and abnormal brain cells. When surgery is unable to completely remove the tumor, radiation therapy becomes a viable option to target the remaining cancer cells and control tumor growth.

Standard External Beam Radiotherapy: Radiation therapy employs a range of radiation beams to precisely target the tumor in a conformal pattern, minimizing exposure to nearby healthy structures. Modern delivery techniques such as three-dimensional conformal radiation (3DCRT) and IMRT further enhance this precision. These methods significantly reduce the risk of long-term radiation damage while effectively treating the tumor [110].

Stereotactic Radiosurgery: Radiation therapy employs advanced devices such as the Gamma Knife, Novalis, and CyberKnife to concentrate radiation from multiple distinct beams onto the target tissue. This targeted approach minimizes damage to surrounding tissues near the tumor. Presently, there is no evidence suggesting the superiority of one delivery method over another in terms of clinical outcomes.

Chemotherapy: Except in cases of atypical or malignant subtypes of meningioma that are not effectively treated with surgery and/or radiation therapy, chemotherapy is seldom used to treat meningioma [111].

Pineoblastoma: Pineoblastoma, a specific type of cancer, originates in the pineal gland located in the middle of the brain. The pineal gland produces melatonin, a hormone that regulates the natural sleep–wake cycle of the body. Pineoblastoma begins with abnormal cell proliferation in the pineal gland, leading to rapid growth and infiltration of healthy tissue. While pineoblastoma can develop at any age, it is more commonly observed in young children [112]. Symptoms of pineoblastoma may include headaches, drowsiness, and abnormalities in eye movement.

Occurrence: Pineal gland tumors are relatively rare, accounting for less than 2% of primary brain tumors. Among these, slightly less than half are pineoblastomas. Pineoblastomas predominantly occur in children and teenagers aged 15 to 40, affecting both genders equally [113].

Symptoms: Any symptoms associated with pineal gland tumors result from the obstruction of CSF flow. Some of these signs include:

Nausea and vomiting

Headache

Double vision

Eye movement problems, such as difficulty looking up [114]

Diagnosis: Pineoblastoma can be detected through various tests and procedures, including the following:

Imaging tests: MRI is commonly used to determine the location and size of the brain tumor. Advanced techniques such as magnetic resonance spectroscopy and perfusion MRI may also be employed. Additional imaging tests such as computerized tomography and positron emission tomography scans may be used for further evaluation.

Biopsy: A biopsy involves the removal of a tissue sample for analysis. This can be performed using a needle before surgery or by extracting the sample during the operation. The tissue sample is examined in a laboratory to identify the types of cells present and their rate of growth [115].

CSF Removal for Testing: A procedure known as a lumbar puncture, or spinal tap, is employed to extract a sample of the fluid surrounding the brain and spinal cord. During this procedure, a medical professional inserts a needle between two lower spine bones to collect CSF surrounding the spinal cord [116]. This fluid is then subjected to testing to detect the presence of pineoblastoma.

Treatment: Treatment options for pineoblastoma may include:

Surgery to remove excess fluid in the brain: In cases where the pineoblastoma obstructs the movement of CSF, leading to increased pressure in the brain, surgery may be performed to alleviate the pressure. During this procedure, a drain or tube may be inserted to redirect the fluid, reducing the pressure. Additionally, a biopsy or surgery to remove the pineoblastoma itself may be performed concurrently [117].

Pineoblastoma removal with surgery: The primary goal of a neurosurgeon is to surgically remove as much of the pineoblastoma as possible. However, complete removal may not always be achievable, especially if the tumor is located close to critical structures deep inside the brain. Additional therapies are often required post-surgery to target any remaining cancer cells [118].

Radiation Therapy: High-energy beams, such as protons or X-rays, are utilized in radiation therapy to target and eliminate cancer cells. During this treatment, beams are directed at the brain and spinal cord using a specialized device. This helps expose cancer cells to higher levels of radiation. Due to the risk of cancer cells spreading from the brain to other areas of the central nervous system, radiation therapy often encompasses the entire brain and spinal cord [119].

3.8 CHEMOTHERAPY

Chemotherapy works by administering potent medications that kill cancer cells. Typically, it is administered following surgery or radiation treatment for pineoblastomas. Occasionally, chemotherapy may be used in combination with radiation therapy. In cases of larger pineoblastomas, chemotherapy may be given before surgery to shrink the tumor, facilitating easier removal [120].

Radiosurgery, employing stereotactic techniques, directs multiple radiation beams precisely at tumor locations to eliminate cancer cells. In instances where pineoblastoma recurs after initial treatment, radiosurgery may be utilized as an effective treatment option.

Clinical trials offer the opportunity to explore cutting-edge therapeutic approaches. However, these medications may have unknown side effects. Patients can discuss the possibility of participating in clinical research studies with their doctor [121].

Radiosurgery: Stereotactic radiosurgery is a specialized technique that directs multiple radiation beams precisely at tumor locations to eliminate cancer cells. In cases where pineoblastoma recurs after initial treatment, radiosurgery may be employed as an effective treatment option.

Clinical trials: Clinical trials provide an opportunity to explore cutting-edge therapeutic approaches. However, these medications may have unknown side effects. Patients are encouraged to inquire with their doctor about the possibility of participating in clinical research studies [121].

REFERENCES

[1] Leo-Kottler B. Brain tumors relevant to clinical neuroophthalmology. In: Schiefer U, Wilhelm H, Hart W, eds. Clinical Neuro-Ophthalmology: A Practical Guide. Berlin: Springer; 2007:171–183.

[2] Tagoe NN, Essuman VA, Fordjuor G, Akpalu J, Bankah P, Ndanu T. Neuro-ophthalmic and clinical characteristics of brain tumors in a tertiary hospital in Ghana. Ghana Med J. 2015;49:181–186.

[3] Rosa RH, Buggage R, Harocopos GJ, et al. Ophthalmic Pathology and Intraocular Tumor, Basic and Clinical Science Course, Section 4. San Francisco, CA: American Academy of Ophthalmology; 2014.

[4] Bhatti M, Biousse V, Bose S, et al. Neuro-Ophthalmology Basic and Clinical Science Course, Section. San Francisco, CA: American Academy of Ophthalmology; 2018.

[5] Sefi-Yurdakul N. Visual findings as primary manifestations in patients with intracranial tumors. Int J Ophthalmol. 2015;8:800–803.

[6] Scheel BI, Holtedahl K. Symptoms, signs, and tests: The general practitioner's comprehensive approach towards a cancer diagnosis. Scand J Prim Health Care. 2015;33:170–177.

[7] Menjot de Champfleur N, Leboucq N, Menjot de Champfleur S, Bonafé A. Imaging of the pre-chiasmatic optic nerve. Diagn Interv Imaging. 2013;94:973–984.

[8] Miller NR. Primary tumors of the optic nerve and its sheath. Eye. 2004;18:1026–1037.

[9] Behbehani R. Clinical approach to optic neuropathies. Clin Ophthalmol. 2007;1:233–246.

[10] Louis DN, Ohgaki H, Wiestler OD, Cavenee WK, eds. World Health Organization Histological Classification of Tumors of the Central Nervous System. Lyon: International Agency for Research on Cancer; 2007.

[11] Jakola AS, Myrmel KS, Kloster R, et al. Comparison of a strategy favoring early surgical resection vs a strategy favoring watchful waiting in low-grade gliomas. J Am Med Assoc. 2012;8(18):1881–1888.

[12] Buckner JC, Pugh SL, Shaw EG, et al. Phase III study of radiation therapy with or without procarbazine, CCNU and vincristine (PCV) in low grade glioma: RTOG 9802 with Alliance, ECOG and SWOG. J Clin Oncol. 2014;32:5s, abstr 2000.

[13] Alford EC Jr, Lofton S. Gliomas of the optic nerve or chiasm. Outcome by patient age, tumor site, and treatment. J Neurosurg. 1988;68:85–98.

[14] Baskin DS, Wilson CB. Surgical management of craniopharyngiomas: A review of 74 cases. J Neurosurg. 1986;65:22–27.

[15] Bird AC, Sanders MD. Choroidal folds in association with papilledema. Br J Ophthalmol. 1973;57:89–97.

[16] Büttner-Ennever JA, Buttner U, Cohen B, et al. Vertical gaze paralysis and the rostral interstitial nucleus of the medial longitudinal fasciculus. Brain. 1982;105:125–149.

[17] Collier J. Nuclear ophthalmoplegia: With especial reference to retraction of lids and ptosis and to lesions of the posterior commissure. Brain. 1927;50:488–498.

[18] Currie JN, Lubin JH, Lessell S. Chronic isolated abducens nerve paresis from tumors at the base of the brain. Arch Neurol. 1983;40:226–229.

[19] Cushing H. The chiasmal syndrome of primary optic atrophy and bitemporal field defects in adults with a normal sella turcica. Arch Ophthalmol. 1930;3:505–551,704–735.

[20] Danesh-Meyer HV, Carroll SC, Foroozan R, et al. Investigative ophthalmology. Vis Sci. 2006;47:4827–4835.

[21] Danesh-Meyer HV, Papchenko TL, Savino PJ, et al. Brightness sensitivity and red desaturation as predictors of relative afferent papillary defect. Invest Ophthalmol Vis Sci. 2008;48:3616–3621.

[22] Danesh-Meyer HV, Papchenko TL, Savino PJ, et al. In vivo retinal nerve fiber layer thickness measured by optical coherence tomography predicts visual recovery after surgery for parachiasmal tumors. Invest Ophthalmol Vis Sci. 2008;49:1879–1885.

[23] Frisen L, Hoyt WF, Tengroth BM. Optociliary veins, disc pallor, and visual loss: A triad of signs indicating spheno-orbital meningiomas. Acta Ophthalmol. 1973;51:241–249.

[24] Glaser JS, Savino PJ, Sumers KD, et al. The photostress recovery test in the clinical assessment of visual function. Am J Ophthalmol. 1977;83:255–260.

[25] Halmagyi GM, Rudge P, Gresty MA, et al. Downbeating nystagmus: A review of 62 cases. Arch Neurol. 1983;40:777–784.

[26] Hoyt WF, Knight CL. Comparison of congenital disc blurring and incipient papilledema in red free light: A photographic study. Invest Ophthalmol. 1973;12:241–247.

[27] Jefferson G. The Bowman Lecture: Concerning injuries, aneurysms and tumors involving the cavernous sinus. Trans Ophthalmol Soc UK. 1953;73:117–152.

[28] Kaye AH. Essential Neurosurgery. Edinburgh: Churchill Livingstone; 1958/1991.

[29] Kearns TP, Rucker CW. Arcuate defects in the visual fields due to chromophobe adenoma of the pituitary gland. I. Clinics in perimetry. Am J Ophthalmol. 1958;45(4 Pt 1):505–507.

[30] King JO. Neuro-ophthalmology of brain tumors. In: Kaye A, Laws ER, eds. Brain Tumors, 2nd ed. London: Churchill Livingstone;2001:249–272.

[31] Kirkham TH. The ocular symptomatology of pituitary tumors. Proc Roy Soc Med. 1972;65:517–518.

[32] Singh AD, Lewis H, Schachat AP. Primary lymphoma of the central nervous system. Ophthalmol Clin. 2005;18:199–207.

[33] Choi JY, Kafkala C, Foster CS. Primary intraocular lymphoma: A review. Semin Ophthalmol. 2006;21:125–133.

[34] Noh T, Walbert T. Chapter 6—brain metastasis: Clinical manifestations, symptom management, and palliative care. In: Schiff D, van den Bent MJ, eds. Metastatic Disease of the Nervous System. Philadelphia, PA: Elsevier; 2018:75–88.

[35] Groom M, Kay MD, Vicinanza-Adami C, Santini R. Optic tract syndrome secondary to metastatic breast cancer. Am J Ophthalmol. 1998;125:115–118.

[36] Zager EL, Hedley-Whyte ET. Metastasis within a pituitary adenoma presenting with bilateral abducens palsies. Neurosurgery. 1987;21:383–386.

[37] Morita A, Meyer FB, Laws ER. Symptomatic pituitary metastases. J Neurosurg. 1998;89:69–73.

[38] Sioutos P, Yen V, Arbit E. Pituitary gland metastases. Ann Surg Oncol. 1996;3:94–99.

[39] Nelson PB, Robinson AG, Martinez JA. Metastatic tumor of the pituitary gland. Neurosurgery. 1987;21:941–944.

[40] Teears RJ, Silverman EM. Clinicopathologic review of 88 cases of carcinoma metastatic to the pituitary gland. Cancer. 1975;36:216–220.

[41] Chiang M, Brock M, Patt S. Pituitary metastases. Neurochirurgia (Stuttg). 1990;33:127–131.

[42] Ebert S, Pilgram SM, Bähr M, Kermer P. Bilateral ophthalmoplegia due to symmetric cavernous sinus metastasis from gastric adenocarcinoma. J Neurol Sci. 2009;279:106–108.

[43] Supler ML, Friedman WA. Acute bilateral ophthalmoplegia secondary to cavernous sinus metastasis: A case report. Neurosurgery. 1992;15:45–46.

[44] Kumar S, Kent SS, Sundaram AN, Sharma M. Unilateral internuclear ophthalmoplegia as an isolated presentation of metastatic melanoma. J Neuroophthalmol. 2015;35:54–56.

[45] Keane JR. Internuclear ophthalmoplegia: Unusual causes in 114 of 410 patients. Arch Neurol. 2005;62:714–717.

[46] Bolaños I, Lozano D, Cantú C. Internuclear ophthalmoplegia: Causes and long-term follow-up in 65 patients. Acta Neurol Scand. 2004;110:161–165.

[47] Wasserstrom WR, Glass JP, Posner JB. Diagnosis and treatment of leptomeningeal metastases from solid tumors: Experience with 90 patients. Cancer. 1982;49:759–772.

[48] Lanfranconi S, Basilico P, Trezzi I, et al. Optic neuritis as isolated manifestation of leptomeningeal carcinomatosis: A case report and systematic review of ocular manifestations of neoplastic meningitis. Neurol Res Int. 2013;2013.

[49] Cantillo R, Jain J, Singhakowinta A, Vaitkevicius VK. Blindness as initial manifestation of meningeal carcinomatosis in breast cancer. Cancer. 1979;44:755–757.

[50] Balm M, Hammack J. Leptomeningeal carcinomatosis: Presenting features and prognostic factors. Arch Neurol. 1996;53:626–632.

[51] Goldenberg-Cohen N, Haber J, Ron Y, et al. Long-term ophthalmological follow-up of children with Parinaud syndrome. Ophthalmic Surg Lasers Imaging. 2010;41:467–471.

[52] Wilkins RH, Brody IA. Parinaud's syndrome. JAMA Neurol. 1972;26:91.

[53] Keane J. The pretectal syndrome: 206 patients. Neurology. 1990;40:684–690.

[54] Villano JL, Propp JM, Porter KR, et al. Malignant pineal germ-cell tumors: An analysis of cases from three tumor registries. Neuro Oncol. 2008;10:121–130.

[55] Hoehn ME, Calderwood J, O'Donnell T, Armstrong GT, Gajjar A. Children with dorsal midbrain syndrome as a result of pineal tumors. J AAPOS. 2017;21:34–38.

[56] Masters PS. The molecular biology of coronaviruses. Adv Virus Res. 2006;65(6):193–292. doi:10.1016/S0065-3527(06)66005-3

[57] Siddell SG, Anderson R, Cavanagh D, et al. Coronaviridae. Intervirology. 1983;20:181–189. doi:10.1159/000149390665464

[58] Virus Taxonomy: 2019 Release. ICTV (International Committee on Taxonomy of Viruses). Available from: https://talk.ictvonline.org/ictv-reports/ictv_9th_report/positive-sense-rna-viruses-2011/w/posrna_viruses/222/coronaviridae

[59] Decaro N, Lorusso A. Novel human coronavirus (SARS-CoV-2): A lesson from animal coronaviruses. Vet Microbiol. 2020;244:108693. doi:10.1016/j.vetmic.2020.10869332402329

[60] Gorbalenya AE, Baker SC, Baric RS, et al. The species severe acute respiratory syndrome-related coronavirus: Classifying 2019-nCoV and naming it SARS-CoV-2. Nat Microbiol. 2020;5(4):536–544. doi:10.1038/s41564-020-0695-z32123347

[61] WHO Coronavirus Disease (COVID-19) Dashboard. Available from: https://covid19.who.int/

[62] Gharebaghi R, Desuatels J, Moshirfar M, Parvizi M, Daryabari SH, Heidary F. Covid-19: Preliminary clinical guidelines for ophthalmology practices. Med Hypothesis Discov Innov Ophthalmol. 2020; 9(2):149–158.

[63] Desforges M, Le Coupanec A, Dubeau P, et al. Human coronaviruses and other respiratory viruses: Underestimated opportunistic pathogens of the central nervous system? Viruses. 2019;12(1):1–28. doi:10.3390/v12010014

[64] Desforges M, Le Coupanec A, Stodola JK, Meessen-Pinard M, Talbot PJ. Human coronaviruses: Viral and cellular factors involved in neuroinvasiveness and neuropathogenesis. Virus Res. 2014;194:145–158. doi:10.1016/j.virusres.2014.09.01125281913

[65] Hamming I, Timens W, Bulthuis MLC, Lely AT, Navis GJ, van Goor H. Tissue distribution of ACE2 protein, the functional receptor for SARS coronavirus. A first step in understanding SARS pathogenesis. J Pathol. 2004;203(2):631–637. doi:10.1002/path.157015141377

[66] Li YC, Bai WZ, Hashikawa T. The neuroinvasive potential of SARS-CoV2 may play a role in the respiratory failure of COVID-19 patients. J Med Virol. 2020;92(6):552–555. doi:10.1002/jmv.2572832104915

[67] Heidary F, Varnaseri M, Gharebaghi R. The potential use of persian herbal medicines against COVID-19 through angiotensin-converting enzyme 2. Arch Clin Infect Dis. 2020;15:e102838. doi:10.1136/bmj.m810

[68] Holappa M, Vapaatalo H, Vaajanen A. Many faces of renin-angiotensin system—focus on eye. Open Ophthalmol J. 2017;11(1):122–142. doi:10.2174/1874364101711101012228761566

[69] Moher D, Liberati A, Tetzlaff J, et al. Preferred reporting items for systematic reviews and meta-analyses: The PRISMA statement. PLoS Med. 2009;6:7. doi:10.1371/journal.pmed.1000097

[70] Wu P, Duan F, Luo C, et al. Characteristics of ocular findings of patients with coronavirus disease 2019 (COVID-19) in Hubei Province, China. JAMA Ophthalmol. 2020;2019:4–7. doi:10.1001/jamaophthalmol.2020.1291

[71] Zhou Y, Duan C, Zeng Y, et al. Ocular findings and proportion with conjunctival SARS-COV-2 in COVID-19 patients. Ophthalmology. 2020. doi:10.1016/j.ophtha.2020.04.028

[72] Hong N, Yu W, Xia J, Shen Y, Yap M, Han W. Evaluation of ocular symptoms and tropism of SARS-CoV-2 in patients confirmed with COVID-19. Acta Ophthalmol. 2020;1–7. doi:10.1111/aos.1444532749776

[73] Navel V, Chiambaretta F, Dutheil F. Haemorrhagic conjunctivitis with pseudomembranous related to SARS-CoV-2. Am J Ophthalmol Case Reports. 2020;19:100735. doi:10.1016/j.ajoc.2020.100735

[74] Xia J, Tong J, Liu M, Shen Y, Guo D. Evaluation of coronavirus in tears and conjunctival secretions of patients with SARS-CoV-2 infection. J Med Virol. 2020;92(6):589–594. doi:10.1002/jmv.2572532100876

[75] Chen L, Liu M, Zhang Z, et al. Ocular manifestations of a hospitalised patient with confirmed 2019 novel coronavirus disease. Br J Ophthalmol. 2020:1–4. doi:10.1136/bjophthalmol-2020-316304

[76] Cheema M, Aghazadeh H, Nazarali S, et al. Keratoconjunctivitis as the initial medical presentation of the novel coronavirus disease 2019. Can J Ophthalmol. 2020;2019:1–5. doi:10.1016/j.jcjo.2020.03.003

[77] Loon SC, Teoh SCB, Oon LLE, et al. The severe acute respiratory syndrome coronavirus in tears. Br J Ophthalmol. 2004;88(7):861–863. doi:10.1136/bjo.2003.03593115205225

[78] Van Der Hoek L, Pyrc K, Jebbink MF, et al. Identification of a new human coronavirus. Nat Med. 2004;10(4):368–373. doi:10.1038/nm102415034574

[79] Daruich A, Martin D, Bremond-Gignac D. Ocular manifestation as first sign of coronavirus disease 2019 (COVID-19): Interest of telemedicine during the pandemic context. J Fr Ophtalmol. 2020;43:389–391. doi:10.1016/j.jfo.2020.04.00232334847

[80] Pei X, Jiao X, Lu D, Qi D, Huang S, Li Z. How to face COVID-19 in ophthalmology practice. Med Hypothesis Discov Innov Ophthalmol. 2020;9(3):164–171.

[81] Sirakaya E, Sahiner M, Sirakaya HAA. A Patient with bilateral conjunctivitis positive for Sars-Cov-2 RNA in conjunctival sample. Cornea. 2020. doi:10.1097/ICO.0000000000002485

[82] Seah I, Agrawal R. Can the coronavirus disease 2019 (COVID-19) affect the eyes? A review of coronaviruses and ocular implications in humans and animals. Ocul Immunol Inflamm. 2020;28(3):391–395. doi: 10.1080/09273948.2020.173850132175797

[83] Marinho PM, Marcos AAA, Romano AC, Nascimento H. Retinal findings in patients with COVID-19. Lancet. 2020;395:1610. doi:10.1016/S0140-6736(20)31014-X32405105

[84] Li Y, Li H, Fan R, et al. Coronavirus infections in the central nervous system and respiratory tract show distinct features in hospitalized children. Intervirology. 2017;59(3):163–169.

[85] Yeh EA, Collins A, Cohen ME, Duffner PK, Faden H. Detection of coronavirus in the central nervous system of a child with acute disseminated encephalomyelitis. Pediatrics. 2004;113(1):e73–e76.

[86] Murray RS, Brown B, Brain D, Cabirac GF. Detection of coronavirus RNA and antigen in multiple sclerosis brain. Ann Neurol. 1992;31(5):525–533. doi:10.1002/ana.4103105111596089

[87] Stewart JN, Mounir S, Talbot PJ. Human coronavirus gene expression in the brains of multiple sclerosis patients. Virology. 1992;191(1):502–505. doi:10.1016/0042-6822(92)90220-J1413524

[88] Wege H, Schluesener H, Meyermann R, BaracLatas V, Suchanek G, Lassmann H. Coronavirus infection and demyelination: Development of inflammatory lesions in Lewis rats. Adv Exp Med Biol. 1998;440:437–444.

[89] Murray RS, MacMillan B, Cabirac G, Burks JS. Detection of coronavirus RNA in CNS tissue of multiple sclerosis and control patients. Coronaviruses and their diseases. Adv Exp Med Biol. 1990;276:505–510. doi:10.1007/978-1-4684-5823-7_70

[90] Bauchet L, Rigau V, Mathieu-Daude H, et al. Clinical epidemiology for childhood primary central nervous system tumors. J Neuro Oncol. 2009;92:87–98.

[91] Ostrom QT, Gittleman H, Farah P, et al. CBTRUS statistical report: Primary brain and central nervous system tumors diagnosed in the United States in 2006–2010. Neuro Oncol. 2013;15(Suppl 2):ii1–56.

[92] Stokland T, Liu JF, Ironside JW, et al. A multivariate analysis of factors determining tumor progression in childhood low-grade glioma: A population-based cohort study (CCLG CNS9702). Neuro Oncol. 2010;12:1257–1268.

[93] Fisher PG, Tihan T, Goldthwaite PT, et al. Outcome analysis of childhood low-grade astrocytomas. Pediatr Blood Cancer. 2008;51:245–250.

[94] Freeman CR, Farmer JP. Pediatric brain stem gliomas: A review. Int J Radiat Oncol Biol Phys. 1998;40:265–271.

[95] Hargrave D, Bartels U, Bouffet E. Diffuse brainstem glioma in children: Critical review of clinical trials. Lancet Oncol. 2006;7:241–248.

[96] Louis DN, Wiestler OD, Cavanee WK, eds. WHO Classification of Tumors of the Central Nervous System. Lyon: International Agency for Research on Cancer; 2007.

[97] Smoll NR. Relative survival of childhood and adult medulloblastomas and primitive neuroectodermal tumors (PNETs). Cancer. 2012;118:1313–1322.

[98] Kleihues P, Burger PC, Scheithauer BW. The new WHO classification of brain tumors. Brain Pathol. 1993;3:255–268.

[99] Rutkowski S, von Hoff K, Emser A, et al. Survival and prognostic factors of early childhood medulloblastoma: An international meta-analysis. J Clin Oncol. 2010;28:4961–4968.

[100] Kool M, Korshunov A, Remke M, et al. Molecular subgroups of medulloblastoma: An international meta-analysis of transcriptome, genetic aberrations, and clinical data of WNT, SHH, Group 3, and Group 4 medulloblastomas. Acta Neuropathol. 2012;123:473–484.

[101] Woehrer A, Slavc I, Waldhoer T, et al. Incidence of atypical teratoid/rhabdoid tumors in children: A population-based study by the Austrian Brain Tumor Registry, 1996–2006. Cancer. 2010;116:5725–5732.

[102] Ostrom QT, Chen Y, Blank PD, et al. The descriptive epidemiology of atypical teratoid/rhabdoid tumors in the United States, 2001–2010. Neuro Oncol. 2014;16:1392–1399.

[103] Lafay-Cousin L, Hawkins C, Carret AS, et al. Central nervous system atypical teratoid rhabdoid tumors: The Canadian Paediatric Brain Tumor Consortium experience. Eur J Cancer. 2012;48:353–359.

[104] .Hilden JM, Meerbaum S, Burger P, et al. Central nervous system atypical teratoid/rhabdoid tumor: Results of therapy in children enrolled in a registry. J Clin Oncol. 2004;22:2877–2884.

[105] von Hoff K, Hinkes B, Dannenmann-Stern E, et al. Frequency, risk-factors and survival of children with atypical teratoid rhabdoid tumors (AT/RT) of the CNS diagnosed between 1988 and 2004, and registered to the German HIT database. Pediatr Blood Cancer. 2011;57:978–985.

[106] Athale UH, Duckworth J, Odame I, Barr R. Childhood atypical teratoid rhabdoid tumor of the central nervous system: A meta-analysis of observational studies. J Pediatr Hematol Oncol. 2009;31:651–663.

[107] Lee JY, Kim IK, Phi JH, et al. Atypical teratoid/rhabdoid tumors: The need for more active therapeutic measures in younger patients. J Neuro Oncol. 2012;107:413–419.

[108} Heck JE, Lombardi CA, Cockburn M, Meyers TJ, Wilhelm M, Ritz B. Epidemiology of rhabdoid tumors of early childhood. Pediatr Blood Cancer. 2013;60:77–81.

[109] Bishop AJ, McDonald MW, Chang AL, Esiashvili N. Infant brain tumors: Incidence, survival, and the role of radiation based on Surveillance, Epidemiology, and End Results (SEER) data. Int J Radiat Oncol Biol Phys. 2012;82:341–347.

[110] Stefanaki K, Alexiou GA, Stefanaki C, Prodromou N. Tumors of central and peripheral nervous system associated with inherited genetic syndromes. Pediatr Neurosurg. 2012;48:271–285.

[111] Hottinger AF, Khakoo Y. Neurooncology of familial cancer syndromes. J Child Neurol. 2009;24: 1526–1535.

[112] Bourdeaut F, Miquel C, Richer W, et al. Rubinstein-Taybi syndrome predisposing to non-WNT, non-SHH, group 3 medulloblastoma. Pediatr Blood Cancer. 2014;61:383–386.

[113] Yu CL, Tucker MA, Abramson DH, et al. Cause-specific mortality in long-term survivors of retinoblastoma. J Natl Cancer Inst. 2009;101:581–591.

[114] Dearlove JV, Fisher PG, Buffler PA. Family history of cancer among children with brain tumors: A critical review. J Pediatr Hematol Oncol. 2008;30:8–14.

[115] Searles Nielsen S, Mueller BA, Preston-Martin S, et al. Family cancer history and risk of brain tumors in children: Results of the SEARCH international brain tumor study. Cancer Causes Control. 2008;19:641–648.

[116] Hemminki K, Kyyronen P, Vaittinen P. Parental age as a risk factor of childhood leukemia and brain cancer in offspring. Epidemiology. 1999;10:271–275.

[117] Yip BH, Pawitan Y, Czene K. Parental age and risk of childhood cancers: A population-based cohort study from Sweden. Int J Epidemiol. 2006;35:1495–1503.

[118] Johnson KJ, Carozza SE, Chow EJ, et al. Parental age and risk of childhood cancer: A pooled analysis. Epidemiology. 2009;20:475–483.

[119] Weinberg CR, Wilcox AJ, Lie RT. A log-linear approach to case-parent-triad data: Assessing effects of disease genes that act either directly or through maternal effects and that may be subject to parental imprinting. Am J Hum Genet. 1998;62:969–978.

[120] Lupo PJ, Nousome D, Okcu MF, Chintagumpala M, Scheurer ME. Maternal variation in EPHX1, a xenobiotic metabolism gene, is associated with childhood medulloblastoma: An exploratory case-parent triad study. Pediatr Hematol Oncol. 2012;29:679–685.

[121] Chen C, Xu T, Chen J, et al. Allergy and risk of glioma: A meta-analysis. Eur J Neurol. 2011;18:387–395.

4 Advanced Imaging Techniques for Brain Tumor

Xiaodong Li¹, Sayantani Garai², Dipro Mukherjee²,
Dibyajit Lahiri², and Moupriya Nag²*
¹Department of Neurosurgery, Shengjing Hospital of China Medical
University, Shenyang, Liaoning, China
²Department of Biotechnology, University of Engineering & Management,
Kolkata, West Bengal, India
*Corresponding author

4.1 INTRODUCTION

The diagnosis and prognosis of cancer pose significant challenges in cancer treatment. Hospitals and medical practitioners utilize various medical imaging techniques of the brain to observe the internal environment and activities without resorting to incisions. Present medical imaging technology broadly employs two main approaches: structural and functional.

Structural imaging techniques are employed to analyze the brain's structure, assess the structural properties of tumor-forming cells, determine tumor location, and identify any other brain injuries or disorders causing structural deformities.

In contrast, functional imaging allows visualization of brain cell activities and monitoring of metabolic changes.

4.2 COMPUTED TOMOGRAPHY IMAGING

The computed tomography (CT) imaging technique utilizes beams of X-rays that rotate around specific parts of the brain. As these X-rays are reflected from various angles, a computer captures the reflected rays to form an image. The X-ray data collected by the computer is used to generate a series of two-dimensional images depicting cross-sectional views of the brain. These images are then combined to produce a three-dimensional (3D) image, offering a more detailed view of the diagnosis of tumors and cancers.

CT scanning has been a staple in clinical diagnosis since the 1970s and has become integral in cerebral studies and tumor evaluation. It is commonly used in cancer diagnosis as it provides detailed images of the internal environment without the need for surgery. However, the use of X-rays in CT scans carries risks, as X-rays emit ionizing radiation, which may increase the risk of cancer, particularly in tissues surrounding the tumor environment.

Despite the emergence of technologies like magnetic resonance imaging (MRI), CT remains a reliable modality due to its ability to detect hemorrhage and calcification and evaluate bone changes caused by tumors. It is particularly preferred in patients with metallic implants or pacemakers, as well as critically ill and unstable patients [1, 2].

4.3 MAGNETIC RESONANCE IMAGING

MRI is preferred over CT scans for brain tumor diagnosis because it produces more detailed pictures [3–5]. MRI utilizes magnetic fields instead of X-rays to create images. During an MRI scan, the magnetic field causes the alignment of protons in the body. When an RF current is pulsed through

DOI: 10.1201/9781003519706-4

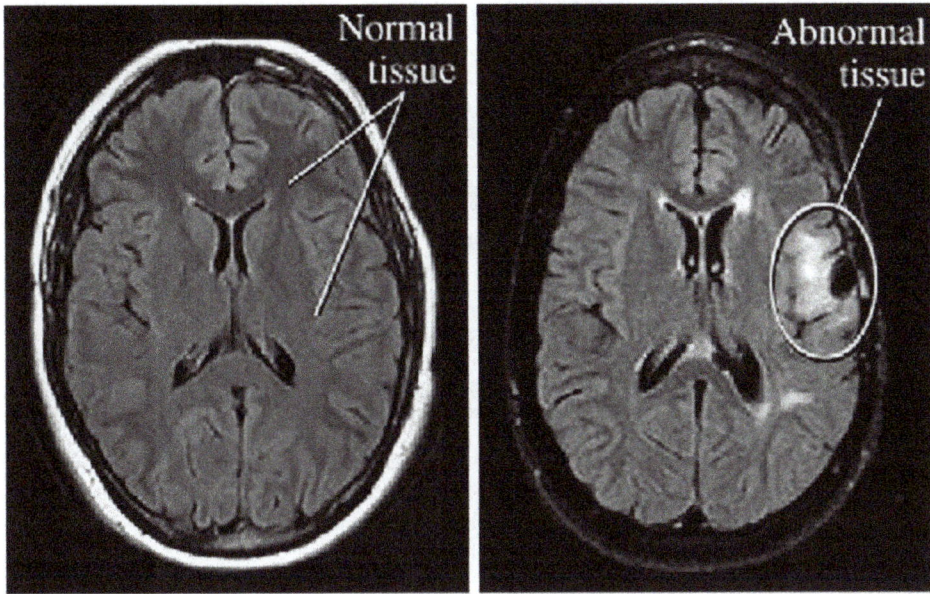

FIGURE 4.1   MRI images showing normal and abnormal brain tissues. The left image depicts a normal brain with labeled normal tissue, while the right image shows an abnormal brain with labeled abnormal tissue, indicating the presence of a tumor.

Source: Anila, S., Sivaraju, S.S., & Devarajan, N. A. New Contourlet Based Multiresolution Approximation for MRI Image Noise Removal. *Natl. Acad. Sci. Lett.* 40, 39–41 (2017). https://doi.org/10.1007/s40009-016-0498-1

the patient, the aligned protons spin out of equilibrium against the magnetic field alignment. External repeated pulses of varying RF energy then perturb the protons, causing realignment with the magnetic field and releasing a certain amount of energy (Figure 4.1).

Detectors and MRI sensors sense this energy, along with the time taken for realignment, and process the data through Fourier transforms. This process results in an arrangement of pixels with varying intensities, producing high-quality images [6].

In MRI, the repetition time (TR) and time to echo (TE) are crucial parameters. TR is the time between successive radio frequency pulses, while TE is the duration between the RF pulse and the emitted signal. MRI sequences can be further characterized by their relaxation times: T1 and T2. T1, or longitudinal relaxation time, measures the time taken by protons to return to equilibrium by realigning with the magnetic field. Conversely, T2, or transverse relaxation time, measures the time taken by excited protons to lose phase coherence [7].

T1-weighted images result from short TR and TE times, enhancing regions of fatty tissues by suppressing signals from water molecules [8]. Another MRI sequencing method, called fluid attenuated inversion recovery (FLAIR), attenuates or enhances the intensity of abnormalities in the cerebrospinal fluid (Figure 4.2). It is used when TR and TE times are very long. T2-weighted images are obtained with longer TR and TE times, enhancing signals from water molecules (Table 4.1).

MRI protocols offer considerable flexibility, allowing clinicians to adjust parameters to obtain high-contrast images. Special dyes or contrast agents, such as gadolinium, are often administered intravenously to patients to enhance proton realignment speed, resulting in brighter T1-weighted images. Gadolinium (Gd) is extensively used in the diagnosis of central nervous system tumors at accepted standard dosages of 0.1 mmol/kg [2].

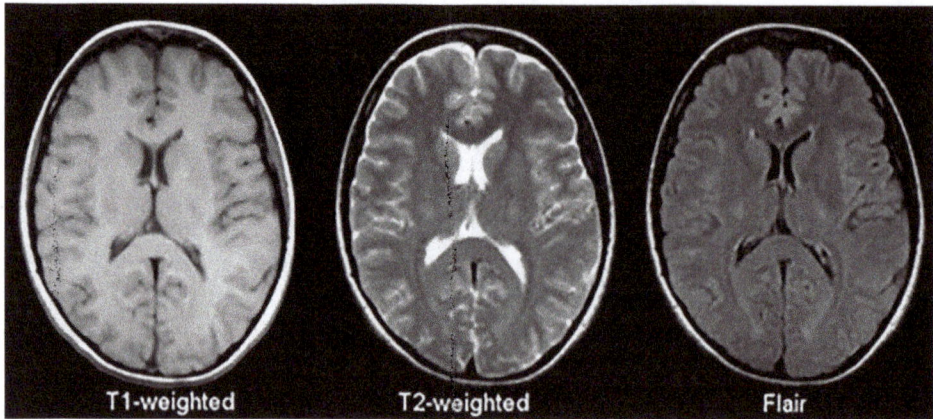

FIGURE 4.2 MRI images illustrating different imaging sequences. From left to right: T1-weighted, T2-weighted, and fluid attenuated inversion recovery (FLAIR) scans. T1-weighted images highlight anatomical details with high-resolution, T2-weighted images emphasize fluid and edema, and FLAIR images are used to suppress cerebrospinal fluid signals to better visualize lesions.

TABLE 4.1
Overview of T1 and T2-Weighted MRI Sequences [7]

	T1-Weighted	T2-Weighted
Characteristics	Short TR and TE	Long TR and TE
TR	~500 ms	~4,000 ms
TE	~14 ms	~90 ms
Signal enhancement	From fatty tissue	From water
Image intensity of specific tissues		
Cerebrospinal fluid	Dark	Bright
White matter	Light	Dark gray
Bone marrow fat	Bright	Light gray
Brain cortex	Gray	Light gray
Inflammation/abnormalities	Dark	Bright

Some studies have suggested that higher doses of the contrast agent can enhance imaging of various intracranial tumors [9–12]. In a similar study, better delineation of tumor lesions and improved visibility of metastases were observed with double or triple doses of Gd [13]. However, increasing the dose of contrast medium carries the risk of toxicity and higher imaging costs for IV Gd-enhanced MRIs.

4.4 FUNCTIONAL MAGNETIC RESONANCE IMAGING

Functional MRI (fMRI) is a functional brain imaging approach that detects and monitors subtle local changes in blood flow in the brain [14]. In a typical protocol, patients are given specific tasks while inside an MRI scanner, and their responses are measured in terms of time-varying local metabolic changes in the brain (Figure 4.3). fMRI scanners utilize this data to produce high-resolution images [15].

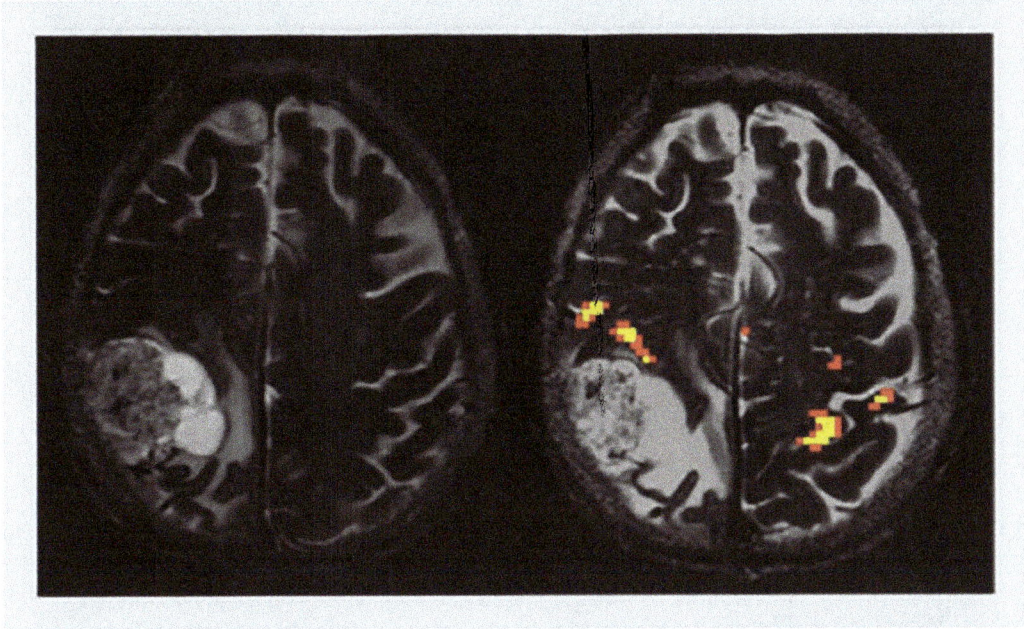

FIGURE 4.3   MRI scans showing brain tumor imaging with functional MRI (fMRI) overlay. The left image displays a conventional MRI scan highlighting a tumor, while the right image includes fMRI data with red and yellow areas indicating regions of increased brain activity around the tumor.

The cognitive tasks performed by patients induce activation or upregulated signaling in certain brain regions, leading to additional neural firing and increased local energy requirement and cerebral oxygen metabolism [16]. This triggers a vasomotor reaction in nearby blood vessels, resulting in dilation and increased blood flow to compensate for the transient oxygen deficit in the surrounding tissues [17]. The resulting hemodynamic response is characterized by an initial surge in deoxygenated hemoglobin (Hb) and a decrease in oxygenated hemoglobin (HbO2), followed by a reversal of these conditions within seconds due to vasodilation [18].

Scanners can easily image the hemodynamic response using blood oxygen level-dependent contrast [19, 20], as both Hb and HbO2 act as endogenous contrast agents due to their paramagnetic and diamagnetic susceptibility, respectively. Alternatively, changes in cerebral blood flow can be visualized, where external agents may be injected to enhance contrast, in a technique known as perfusion-weighted MRI [21].

4.5   POSITRON EMISSION TOMOGRAPHY

Positron emission tomography, as the name suggests, is a technique that measures and quantifies emitted positrons or positive electrons to create images. In the usual protocol radioactive tracer is injected intravenously to measure parameters like neuro transmittance, metabolism of tissues, or blood flow rates (Figure 4.4). PET can efficiently produce bright and colorful images that can distinguish the differential blood flow levels and oxygen consumption its different parts of the brain or even track neurotransmitters such as dopamine and therefore is often used as a functional modality of brain imaging [22]. Since malignant brain tumors are known to be associated with increased glucose metabolism, tracers such as fluorodeoxyglucose can be used to quantify and create images that clearly distinguish cancerous tissues as brighter pixels of high intensity [23, 24].

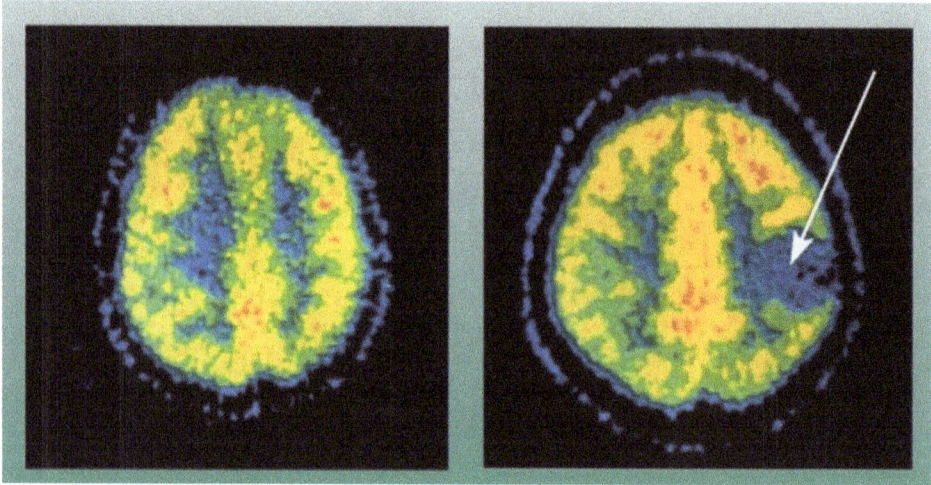

FIGURE 4.4 PET scans illustrating metabolic activity in the brain. The left image shows a normal brain with evenly distributed metabolic activity. The right image indicates abnormal metabolic activity with an arrow pointing to a tumor, where there is increased uptake of the radioactive tracer, indicating higher metabolic rates typical of cancerous tissues.

4.6 CONCLUSION

Imaging techniques like CT and MRI have long been employed for the detection of brain tumors. However, advancements in technology and ongoing research have led to the development of new modalities such as PET, SPECT, and diffusion tensor imaging. These modalities offer additional insights into and capabilities for diagnosing brain cancer, where early detection is crucial for effective treatment (Figures 4.5 and 4.6).

Each imaging modality has its own advantages and limitations. Therefore, a combination of different modalities is often utilized for comprehensive diagnosis and confirmation of brain tumors. These imaging techniques, when combined with machine learning classification-based algorithms, have the potential to revolutionize diagnostics and treatment approaches.

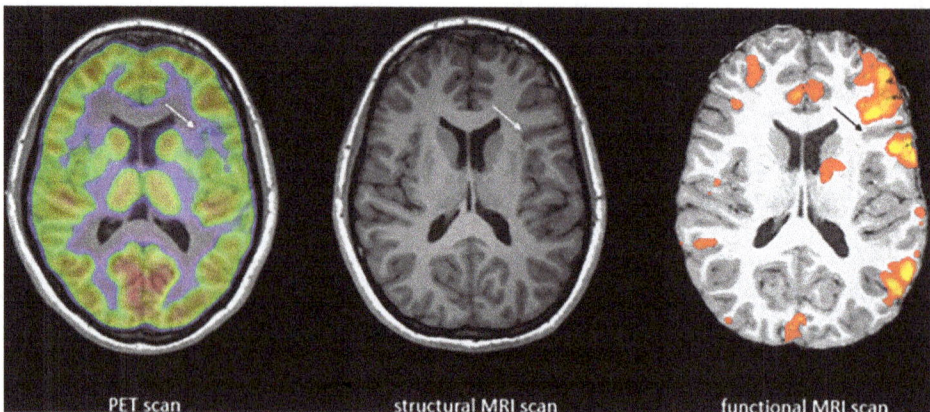

FIGURE 4.5 Comparison of PET, structural MRI, and functional MRI scans. The PET scan (left) shows metabolic activity with varying color intensities. The structural MRI scan (center) provides detailed anatomical information. The functional MRI scan (right) highlights areas of increased brain activity, indicated by red and yellow regions.

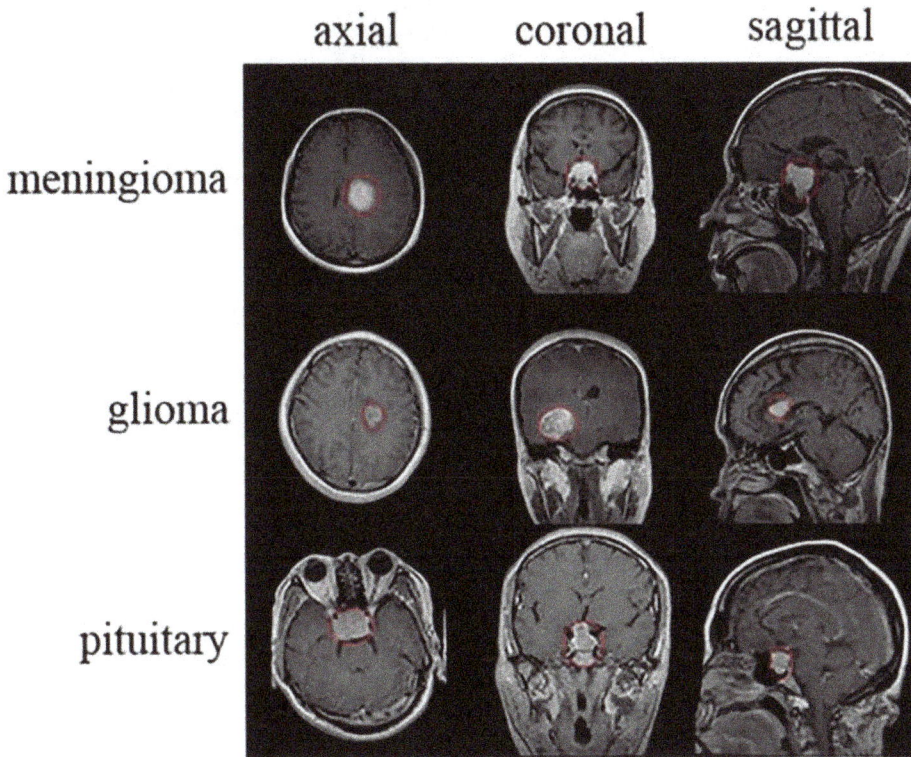

FIGURE 4.6 MRI scans of different brain tumors in axial, coronal, and sagittal planes. The top row shows meningioma, the middle row shows glioma, and the bottom row shows a pituitary tumor. Each tumor type (enclosed in red circles) is displayed in three different planes: axial (left column), coronal (middle column), and sagittal (right column), highlighting the varied perspectives used in imaging and diagnosing brain tumors.

By integrating advanced imaging technologies with machine learning algorithms, healthcare professionals can enhance the accuracy and efficiency of brain tumor diagnosis, leading to improved patient outcomes.

REFERENCES

[1] Whelan, H. T., Clanton, J. A., Wilson, R. E., et al. (1988). Comparison of CT and MRI brain tumor imaging using a canine glioma model. *Pediatric Neurology*, 4(5), 279–283.

[2] Drevelegas, A., & Papanikolaou, N. (2010). Imaging modalities in brain tumors. *Imaging of Brain Tumors with Histological Correlations*, 13–33. https://doi.org/10.1007/978-3-540-87650-2_2

[3] Maravilla, K., & Sory, W. (1986). Magnetic resonance imaging of brain tumors. *Seminars in Neurology*, 6(1), 33–42. https://doi.org/10.1055/s-2008-1041445

[4] Brant-Zawadzki, M., Badami, J. P., Chi, M., et al. (1984). Primary intracranial tumor imaging: A comparison of magnetic resonance and CT. *Radiology*, 150, 435–440.

[5] Bradley, W. G., Waluch, V., Yadley, R. A., et al. (1984). Comparison of'CT and MR in 400 patients with suspected disease of the brain and cervical spinal cord. *Radiology*, 152, 695–702.

[6] Magnetic Resonance Imaging (MRI). (2023). National Institute of Biomedical Imaging and Bioengineering. www.nibib.nih.gov/science-education/science-topics/magnetic-resonance-imaging-mri

[7] MRI Basics. (2023). Case.edu. https://case.edu/med/neurology/NR/MRI%20Basics.htm

[8] Kawahara, D., & Nagata, Y. (2021). T1-weighted and T2-weighted MRI image synthesis with convolutional generative adversarial networks. *Reports of Practical Oncology and Radiotherapy*, 26(1), 35–42. https://doi.org/10.5603/rpor.a2021.0005

[9] Runge, V. M., Kirsch, J. E., Burke, V. J., et al. (1992). High dose gadoteridol in MR imaging of intracranial neoplasm. *Journal of Magnetic Resonance Imaging*, 2, 9–18.

[10] Yuh, W. T., Fisher, D. J., Runge, V. M., et al. (1994). Phase III multicenter trial of high-dose gadoteridol in MR evaluation of brain metastases. *American Journal of Neuroradiology*, 15, 1037–1051.

[11] Yuh, W. T., Fisher, D. J., Engelken, J. D., et al. (1991). MR evaluation of CNS tumors: Dose comparison study with gadopentate dimeglumine and gatoteridol. *Radiology*, 180, 485–491.

[12] Yuh, W. T., Nguyen, H. D., Tali, E. T., et al. (1994). Delineation of gliomas with various doses of MR contrast material. *American Journal of Neuroradiology*, 15, 983–989.

[13] Van Dijk, P., Sijens, P. E., Schmitz, P. I. M., et al. (1997). Gd-enhanced MR imaging of brain metastases: Contrast as a function of dose and lesion size. *Magnetic Resonance Imaging*, 15, 535–541.

[14] Rsna, A. (2022). Functional MRI (fMRI). *Radiologyinfo.org*. www.radiologyinfo.org/en/info/fmribrain

[15] Glover, G. H. (2011). Overview of functional magnetic resonance imaging. *Neurosurgery Clinics of North America*, 22(2), 133–139. https://doi.org/10.1016/j.nec.2010.11.001

[16] Buxton, R., & Frank, L. (1997). A model for the coupling between cerebral blood flow and oxyen metabolism during neural stimulation. *Journal of Cerebral Blood Flow & Metabolism*, 17, 64.

[17] Buxton, R. B., Wong, E. C., & Frank, L. R. (1998). Dynamics of blood flow and oxygenation changes during brain activation: The balloon model. *Magnetic Resonance in Medicine*, 39, 855.

[18] Davis, T. L., Kwong, K. K., Weisskoff, R. M., et al. (1998). Calibrated functional MRI: Mapping the dynamics of oxidative metabolism. *Proceedings of the National Academy of Sciences of the United States of America*, 95, 1834.

[19] Ogawa, S., Lee, T. M., Kay, A. R., et al. (1990). Brain magnetic resonance imaging with contrast dependent on blood oxygenation. *Proceedings of the National Academy of Sciences of the United States of America*, 87, 9868.

[20] Ogawa, S., Menon, R. S., Tank, D. W., et al. (1993). Functional brain mapping by blood oxygenation level-dependent contrast magnetic resonance imaging. A comparison of signal characteristics with a biophysical model. *Biophysical Journal*, 64, 803.

[21] Belliveau, J. W., Kennedy, D. J., McKinstry, R. C., et al. (1991). Functional mapping of the human visual cortex by magnetic resonance imaging. *Science*, 254, 716.

[22] Berger, A. (2003). Positron emission tomography. *BMJ*, 326(7404), 1449. https://doi.org/10.1136/bmj.326.7404.1449

[23] Hofman, M. S., & Hicks, R. J. (2016). How we read oncologic FDG PET/CT. *Cancer Imaging*, 16(1). https://doi.org/10.1186/s40644-016-0091-3

[24] Wong, T. Z., van der Westhuizen, G. J., & Coleman, R. E. (2002). Positron emission tomography imaging of brain tumors. *Neuroimaging Clinics of North America*, 12(4), 615–626. https://doi.org/10.1016/s1052-5149(02)00033-3

5 Radiosurgery for Brain Tumors

Xiaodong Li[1], Dipro Mukherjee[2], Sayantani Garai[2],
Mansi Agarwal[3], Sanchari Das[2], Dibyajit Lahiri[2],
and Moupriya Nag[2]*

[1]Department of Neurosurgery, Shengjing Hospital of China Medical University, Shenyang, Liaoning, China
[2]Department of Biotechnology, University of Engineering & Management, Kolkata, West Bengal, India
[3]Department of Bioscience & Bioengineering, Indian Institute of Technology, Jodhpur
*Corresponding author

5.1 INTRODUCTION

The principles that work to make conventional radiation effective are as follows: repair and repopulation increase tumor cell survival, while reassortment, reoxygenation, and radiosensitivity enhance tumor cell kill [1]. However, when considering brain cancer treatment, the method of radiation administration can yield both benefits and drawbacks. Some of these drawbacks may be inherent, while others are speculative.

For instance, stereotactic radiosurgery (SRS) or any hypo-fractionated regimen may face a radiobiological drawback, such as the inability to capitalize on the redistribution of the cell cycle [1]. In comparison to a lengthy conventional radiation regimen, single-fraction SRS might not always achieve optimal cell killing, as cells may not have enough time to redistribute into the G2 and M phases of the cell cycle, which are more radiosensitive [2].

Traditional radiation offers another benefit: the ability to utilize reoxygenation in hypoxic tumor cells, which is considered a biological drawback of single-fraction SRS. However, SRS may have a radiobiological benefit over standard radiotherapy by reducing the repopulation of tumor cells [3]. Additionally, single-fraction SRS might enhance anti-tumor immunity following tumor irradiation, a phenomenon known as the abscopal effect.

According to numerous clinical trials involving melanoma, administering hypo-fractionated radiation or SRS at a specific tumor site may trigger the anti-tumor immune system to reject metastatic lesions at other locations [4]. This phenomenon, known as the abscopal effect, was observed when ipilimumab, a monoclonal antibody, was combined with radiation treatment. Another example occurred in a patient with metastatic lung adenocarcinoma who received a hypo-fractionated radiation regimen alongside ipilimumab, resulting in a complete response at both the primary and distant tumor sites [5].

Determining which of these principles is most critical for a particular tumor is uncertain, as clinical results often defy established biological principles. However, several clinical studies support SRS as an effective therapeutic option for various conditions, including functional impairments, benign and malignant brain tumors, and arteriovenous malformations (AVMs) [6].

The application of the linear-quadratic (LQ) model, which considers repair, repopulation, reassortment, reoxygenation, and radiosensitivity, is common in conventional radiotherapy. The primary difference between SRS and conventional radiotherapy lies in the dosage and target volume. SRS delivers high doses of precisely targeted radiation with a steep dose fall-off, often in a single procedure [7].

The alpha/beta (a/b) ratio, which determines the response of tissues to radiation, is crucial in calculating the dose for tumor control while limiting damage to normal tissues. Malignant tumors

DOI: 10.1201/9781003519706-5

typically have higher a/b ratios, indicating quick tissue response, while benign tumors have lower ratios, indicating slower response [8].

However, applying the LQ model to SRS remains controversial. Some argue that the model underestimates tumor control at the high doses used in SRS and fails to account for other mechanisms of tumor cell killing, such as vascular damage [9]. Alternative models, like the lethal-potentially lethal model or modified LQ model, have been proposed to address these limitations [10].

While there is disagreement about the applicability of the LQ model to SRS, it is clear that SRS induces DNA damage and potentially other mechanisms of tumor cell death. Ongoing research aims to refine our understanding of the radiobiology of SRS and improve treatment outcomes.

5.2 CRANIAL NERVES: BIOLOGICAL EFFECTS

SRS is not only used to treat brain tumors but also functional conditions like trigeminal neuralgia, with primate models being employed to study its effects. In these models, each proximal nerve anterior to the pons received a single 4 mm isocenter with a maximum dosage of either 80 or 100 Gy. Histological examination using both optical and electron microscopy revealed distinct effects compared to a placebo group. Nerves exposed to 80 Gy showed focal axonal degradation and mild edema, while those exposed to 100 Gy exhibited nerve necrosis [11]. These findings were corroborated by studies in rhesus monkeys, which showed minimal impact on trigeminal nerve structure at 60 and 70 Gy, but demyelination at 80 Gy followed by neuronal necrosis at 100 Gy [12].

5.3 BLOOD VESSELS

SRS is believed to induce a proliferative vasculopathy in the arteries of AVMs, with similar effects observed in benign tumors. This vasculopathy begins with damage to endothelial cells due to exposure to high levels of ionizing radiation [13]. Following endothelial cell damage, veins undergo hyalinization and thickening, leading to luminal constriction. With persistent reaction, a significant portion of the AVM may be replaced by gliotic scarring or fibroblasts. Myofibroblasts contribute to wall shrinkage, ultimately obliterating the AVM [14]. Additionally, along with induced thrombosis, within two to three years, 75% of AVM vasculature is completely destroyed [15].

In recent preclinical studies using rat models, the development of AVMs has been linked to the Notch1 and Notch4 signaling pathways. When SRS was applied alongside a dose of 25 Gy, increased apoptotic activity and reduced expression of Notch receptors were observed. In a rat model of AVMs, this resulted in the constriction of the nidus vessels and thrombotic occlusion [16]. The clinical outcomes related to treating blood vessel abnormalities have been extensively documented [17].

SRS is a frequently used therapy strategy for patients suffering from AVMs in the therapeutic setting. When treated with microsurgery, SRS can cause substantial and undesirable rates of death and morbidity in individuals having severe or Spetzler-Martin grades IV and V AVMs.

5.4 BENIGN TUMOR STEREOTACTIC RADIOSURGERY

SRS is a common treatment for benign tumors, with documented results for almost all types of intracranial tumors. However, due to the high survival rate (95%) among treated individuals and the limited number of tumor removals, there is a lack of tissue available for histological analysis to assess the effects of SRS on certain tumor types. An experiment aimed to compare histopathological changes in radiation-induced regions, delineated by imaging enhancement, to those without enhancement. In individuals with WHO grade I meningioma, biopsies from enhancing regions revealed inflammation, demyelination, and cystic alterations [18]. Biopsies from sites lacking enhancement showed coagulation necrosis, edema, vasculopathy, and reactive astrocytosis. Similar pathological alterations, including radiobiological changes such as cytotoxic and delayed vascular effects, were observed in the removal of schwannomas, pituitary tumors, and many other benign neoplasms.

A study evaluating the impact on benign tumors, including schwannomas, employed a xenograft experiment using nude mice hosting human vestibular schwannomas (VSs) in the subrenal capsule [19]. Tumor growth and vascularity were assessed following SRS with single doses of 10, 20, or 40 Gy. Histological analysis revealed significantly reduced tumor growth at 20 and 40 Gy, along with vascular alterations characterized by increased hemosiderin accumulation and mural hyalinization.

To investigate the in vivo radiobiology of acoustic schwannomas post-SRS, a subrenal capsule xenograft in nude mice served as a valuable model [20]. These preclinical findings shed light on how SRS affects the cell damage of benign tumors. In a clinical setting, an MRI-based stereotactic strategic plan enables accurate visualization of the tumor and surrounding brain tissue. Using numerous tiny irradiation isocenters reduces therapeutic adverse effects such as cranial nerve neuropathy or brainstem impacts while enabling conformal SRS to cranial base malignancies [21]. The SRS strategy should be intensely focused and responsive to the target. The following section will highlight the importance of cochlear dosage limitations.

5.5 MALIGNANT TUMOR STEREOTACTIC RADIOSURGERY

Considerable research has been dedicated to malignant tumor SRS, especially primary and metastatic brain tumors. While in brain metastasis, SRS is the most common diagnosis at various locations, several experimental studies have utilized glioma models. Furthermore, the responsiveness of glioma cell lines to different radiation dosages has been employed as a model to understand the responses to SRS [22].

In experiments utilizing an in vivo rat malignant glioma model, single-fraction, localized radiation has demonstrated tumoricidal and cytotoxic effects [23]. Preclinical research on these rat models, with C6 glioma cells implanted in the right frontal brain region, investigated the effects of SRS and standard radiation on histological alterations. Rats were randomly assigned to placebo or treatment groups receiving various radiation treatments. These treatments included SRS to 35 Gy at the tumor site (n = 22), 20 Gy in 5 fractions of whole-brain radiation (n = 18), SRS with whole brain radiation (n = 13), 85 Gy in 10 fractions of partial-brain radiation (n = 16), or 35 Gy in 1 fraction of single-fraction partial-brain radiation (n = 10). Overall survival (OS) significantly increased in all groups receiving radiation treatments compared to the control group (median survival of 22 days). Except for those receiving whole-brain radiation, all treated groups exhibited reduced tumor size and improved survival rates. Notably, animals receiving SRS with SRS plus whole-brain radiation therapy (WBRT) showed decreased tumor cell density and enhanced intratumoral edema compared to the placebo group, a phenomenon not observed with other treatment strategies. Histological responses following SRS were typically more pronounced than those obtained with biologically similar doses of standard radiation therapy, indicating increased tumor cytotoxicity [24].

From a biological standpoint, noteworthy considerations include bevacizumab's potential for synergistic effects and the radiosurgical targeting of white matter tumor cell invasion [25].

5.6 BENIGN TUMORS

5.6.1 Vestibular Schwannomas

As SRS has evolved, management strategies for VSs have undergone significant transformations. Initially, microsurgery was the standard procedure for all patients able to tolerate surgery, while SRS was primarily reserved for individuals with neurofibromatosis type II or those deemed unsuitable for surgery due to age or other factors [26]. However, recent studies indicate a notable shift in practice patterns, with a 41% decrease in surgery rates for VSs observed.

When managing VSs, several considerations come into play, including controlling tumor growth, preventing facial and trigeminal nerve dysfunction, preserving functional hearing, maintaining or improving overall neurological function, and reducing comorbidities such as infection, hydrocephalus, and hemorrhage. Over the years, SRS procedures have been refined to better balance tumor management and comorbidity prevention.

Significant advancements in SRS approaches include the transition from CT to MRI-based modeling, improved computer workstations, precise dose planning, the use of additional radiation isocenters, and narrower irradiation beams. The most common tumor margin dose has become 12 and 13 Gy, which can be customized for each patient based on cranial nerve function, tumor volume, and clinical history. These advancements have led to a significant reduction in the morbidity associated with SRS [27].

Preserving hearing is a priority for many individuals undergoing SRS for VSs. Recent research highlights the significant impact of radiation dosage on the cochlea on hearing preservation outcomes. Strategies aimed at reducing cochlear dosage below 4 Gy tend to yield higher rates of functional hearing retention [28]. Techniques such as beam blocking and the use of small collimators are employed to achieve this goal. However, confining the dose to the cochlea can be challenging in cases where tumors extend into the internal auditory canal. Two techniques to address this challenge involve reducing the marginal dosage and limiting treatment to the lateral aspect of the tumor [29]. While standard radiation therapy may control maximum cochlear dosage, it has not been shown to increase hearing preservation rates [30].

There is ongoing debate regarding the significance of cochlear dosage in SRS treatment planning. Some authors argue that tumor control should take priority over cochlear protection unless lower dose concentrations are proven effective in controlling tumor growth [31]. Additionally, the patient's overall cochlear volume may influence hearing outcomes, allowing for more aggressive treatment approaches in patients with higher hearing reserves [32].

Studies indicate that patients with usable hearing prior to SRS can maintain functional hearing rates of 60% to 85% in the short term [33]. The probability of hearing retention in the near term for patients with intracanalicular tumors is over 80% [34]. However, long-term studies show a gradual decline in functional hearing, with rates ranging from 23% to 80% at ten years post-SRS [35, 36]. Prognostic markers for hearing impairment include pre-treatment hearing ability and tumor size.

A recent study comparing the rate of hearing loss after SRS to the natural course of hearing loss in the VS population found that SRS resulted in a lower rate of hearing loss (3.77 dB/year) compared to the natural course (5.39 dB/year) [37]. The study also identified a cochlear dosage of 4 Gy as effective in preserving hearing. Moreover, research on SRS for VS in individuals over 40 years of age indicates that 93% of patients maintain functional hearing 10 years post-SRS [38].

In contrast, microsurgical studies report functional hearing preservation rates of 30% to 60% initially, with 25% of patients retaining functional hearing in long-term follow-up after retrosigmoid microsurgical resection [39].

Preserving facial nerve functionality is another key objective of VS therapy, with SRS demonstrating benefits in achieving this goal. Studies show a minimal risk of delayed facial nerve damage in patients receiving doses below 13 Gy, with some patients even experiencing improved facial nerve functionality post-SRS [40]. While the likelihood of long-term SRS complications is low, transient issues such as trigeminal nerve impairment and hydrocephalus may occur. Malignant transformation of VS after SRS is exceedingly rare, with only a few reported cases to date [41].

Patients who are not suitable candidates for surgery or prefer to delay intervention may opt for conservative management of VS. Studies indicate that between 70% and 95% of patients choosing this approach experience measurable tumor growth after five and ten years, respectively. In contrast, patients undergoing SRS typically have tumor control rates ranging from 93% to 100% [42].

At one, two, and five years, research showed that the percentage of patients retaining functional hearing through observation decreased to 78%, 43%, and 14%, respectively [43]. In contrast,

SRS demonstrated functional hearing rates ranging from 50% to 70% at the five-year follow-up. Comparatively, observation often leads to tumor growth and a decline in functional hearing. Most patients prefer SRS over microsurgery because it offers similar tumor control rates and better retention of hearing and facial nerve function without the side effects of invasive treatment. Traditionally, managing large VSs has required resection. A study of 246 patients with brainstem compression over 17 years found a 97% tumor control rate with preserved functional brain tissue. Another study examined 65 individuals with VSs ranging from 3 to 4 cm in length and found that 89% of the VSs remained stable or decreased in size at two years, with only 12% requiring additional treatment. Similarly, in a trial involving 28 patients with VSs measuring 3 to 4 cm, 86% retained facial nerve and hearing levels after two years of tumor treatment, with only one patient experiencing a lasting issue despite an 80% transient complication rate.

Additionally, a study of 24 patients with VSs larger than 3 cm demonstrated progression-free survival rates of around 95% and 82% at three and five years, respectively. These findings suggest that SRS remains a viable option for some large VSs, especially in patients who are not suitable candidates for surgery [44]. SRS for VS capitalizes on the hypothesis that normal tissue can repair itself between fractions, potentially impacting the target and surrounding tissue biologically. Large, asymptomatic lesions near the vestibular or trigeminal systems are expected to benefit the most from this approach. The Stanford group reported a crude percentage of usable hearing retention of 76%, no facial weakness, and a five-year tumor control rate of 96% [45]. Several other studies have reported tumor control rates ranging from 94% to 95%. Significant research has shown hearing retention rates of 57% at two years, 71% at five years, and 61% at ten years [46]. A recent study characterized the use of conventional radiotherapy (46.8 Gy in 1.8 Gy fractions) and demonstrated that 54% hearing retention at five years could be achieved without compromising tumor control rates [47]. However, while improving hearing outcomes is a primary goal of fractionated dosing, overall, hearing outcomes after fractionated SRS are not superior to those after single-fraction SRS.

5.7 MENINGIOMAS

SRS has demonstrated remarkable success in treating patients with benign (WHO grade I) meningiomas, which are typically well-defined and occasionally infiltrative. With meticulous planning, it's possible to deliver a high dose of radiation while minimizing exposure to nearby vital tissues. Initially, SRS was primarily considered for treating residual or recurrent meningiomas after surgical intervention [48]. However, concerns regarding delayed tumor recurrence post-surgery, surgical morbidity, and mortality, especially in older patients, have led to a shift in perspective. Now, more practitioners view SRS as the primary mode of tumor management, particularly when achieving a curative grade 1 resection is challenging [49].

In numerous studies, the long-term control rate for benign meningiomas exceeds 90%. For instance, a comprehensive investigation conducted by the University of Pittsburgh included 1,045 intracranial meningiomas among 972 patients. The cohort comprised predominantly women (70%), with nearly half having undergone prior resection and a small percentage (5%) having received standard radiation therapy. Tumors were located in various regions, with an average volume of 7.4 mL. Follow-up periods ranged from 5 to 12 years for different subsets of patients [50].

Presently, postoperative SRS is recommended over leaving residual gross tumor to reduce the risk of delayed progression, rather than opting for subtotal resection (STR) followed by observation. This approach is particularly advantageous for patients under 75 years old. Longitudinal studies have shown that untreated meningiomas under observation tend to increase in size over time. Therefore, researchers believe that SRS is the preferred strategy, especially for young patients with critical condition small meningiomas. For many individuals, surveillance is no longer considered the optimal approach, especially if they are experiencing symptoms.

5.8 PITUITARY ADENOMAS

Another alternative in this scenario is traditional radiation or fractionated radiotherapy. For example, one trial utilizing the CyberKnife fractionated irradiation system demonstrated stabilization of imaging-defined tumors and preservation or improvement of vision in all 20 patients at a 30-month follow-up [51]. Pituitary adenoma SRS aims to preserve pituitary functionality, restore hormone production in functioning adenomas, and safeguard neurological function, particularly vision [51]. Relapse or continued growth of pituitary adenomas after excision are typical indications for SRS.

Recent research outcomes have been categorized by the type and location of meningioma. An analysis of 115 individuals with convexity meningiomas treated with a mean tumor margin dose of 14.2 Gy revealed actuarial tumor control rates of 95% and 86% at three and five years, respectively, for patients with benign meningiomas who had not undergone prior surgery. The cumulative morbidity rate was 10%, and 5% of patients exhibited symptomatic peritumoral imaging abnormalities consistent with edema or adverse radiation effects [52].

5.9 BRAIN METASTASES

An increasing number of studies focusing on the therapeutic applications of SRS have coincided with the growing significance of SRS in managing brain metastases. For many years, WBRT has been the standard approach for treating single or multiple brain metastases. However, SRS has emerged as a preferred method for well-defined metastatic brain tumors due to its ability to provide precise targeting while minimizing radiation exposure to healthy brain tissue through its inherent dose fall-off properties. Initially, SRS was administered alongside WBRT, but it is now increasingly used alone to treat solid tumors, irradiate postoperative tumor beds, or in combination with systemic biologic, cytotoxic, or immune therapies with access to the CNS [53]. The goal of SRS is to achieve brain control in patients with brain metastases while reducing the risk of long-term neurotoxic or cognitive side effects associated with WBRT, either with or without delayed WBRT [54]. Common side effects of WBRT include hair loss, fatigue, and subacute as well as delayed cognitive abnormalities, which can significantly impact the quality of life of long-term survivors [55]. To minimize these toxicities, techniques such as intensity-modulated radiation therapy, hippocampus-specific dose reduction (as investigated in a current Radiation Therapy Oncology Group [RTOG] trial), and the use of neuroprotectors have been employed [56]. However, SRS is increasingly favored as a complement or alternative to WBRT due to its minimally invasive nature, outpatient-based procedure, short recovery time, superior local control, symptom alleviation, and avoidance of craniotomy. Technological advancements in imaging (primarily high-resolution MR imaging) and SRS delivery technologies have enabled the identification and treatment of numerous tumors in suitable patients.

Long-term research findings suggest that the toxicity of SRS is generally minimal, with the exception of a small risk of harm to vital organs, which can typically be managed medically without significantly impacting the patient's quality of life [57].

5.10 GLIAL NEOPLASMS

Adults with primary brain tumors most commonly present with glioblastoma multiforme (GBM). Unfortunately, one of the most significant challenges in neuro-oncology remains the management and treatment of malignant gliomas (WHO grades III and IV). The current treatment protocol typically involves maximal safe surgical resection, followed by standard radiation therapy along with concurrent and adjuvant temozolomide [58]. Despite these aggressive treatments, nearly all patients experience local recurrence and tumor progression, necessitating further therapy, often including additional chemotherapy or SRS [59].

The debate over increasing the dosage of conventional radiotherapy to reduce the rate of local recurrence versus employing SRS to target recurrent lesions while minimizing exposure to surrounding brain parenchyma is ongoing. While some advocate for higher doses of traditional radiotherapy to address local recurrence, many favor the use of SRS due to its ability to precisely target recurrent lesions. However, questions remain regarding the optimal dosage increase and whether SRS is more effective when combined with other treatments. These topics continue to be subjects of active discussion and research in the field.

Research on the role of SRS in low-grade gliomas (WHO grades I and II) is notably limited. Given evidence indicating that the extent of surgical resection correlates with patient survival, these tumors are often treated with maximal surgical resection followed by standard radiation therapy, when feasible [60]. Consequently, SRS is usually reserved for patients with smaller, low-grade gliomas located in challenging areas such as the brainstem, deep brain structures, and critical functional cortex, as well as for cases of residual or recurrent tumors [61].

SRS plays a crucial role in the treatment of glial neoplasms, particularly in malignant gliomas. It is often utilized as a radiation boost following conventional chemoradiation to prevent local recurrence or when the disease shows progression. While additional research is necessary to evaluate the safety and efficacy of this combination compared to other salvage therapies, there is increasing interest in using SRS alongside bevacizumab. In the case of low-grade gliomas, SRS is commonly employed for recurrent tumors in both adults and children, as well as for tumors located in critical areas where surgical excision is not feasible.

5.11   EPENDYMOMAS

Ependymomas account for roughly 6% to 12% of intracranial tumors in children and 5% in adults. Standard treatment typically involves surgical resection followed by radiation therapy. However, the extent of surgical resection is a critical prognostic factor, with studies consistently showing that gross residual tumor following surgery is associated with a poorer prognosis. Unfortunately, achieving gross total resection (GTR) is not always feasible in clinical practice.

There is limited documentation on the use of SRS in ependymomas in the research literature. Several small studies with varying outcomes have been reported. The longest series conducted to date involved a total of 39 individuals with a median age of 22.8 years and 56 tumors who received SRS treatment. Most patients had previously undergone standard radiotherapy following surgical removal of their tumors. The median marginal dosage was 15 Gy, with median doses for WHO grade II and III ependymomas at 13 and 16 Gy, respectively.

Following SRS treatment, the median patient survival was 19.4 months, with a 73.2% local tumor control rate at a median follow-up of 14.9 months. Progression-free survival rates after one and five years were 81.5% and 45.8%, respectively. Subsequent studies indicated that spinal spread was associated with poorer OS, while smaller SRS target volumes and uniform contrast enhancement on MRI were associated with improved progression-free survival. Patient age and tumor grade based on histology did not significantly affect progression-free survival.

Adverse radiation effects were observed in 3 out of 39 patients (7.7%), with two of them requiring treatment with oral corticosteroids. One patient exhibited significant contrast enhancement and peritumoral T2 signal alterations on MRI, although they were asymptomatic. A later stereotactic biopsy revealed the presence of a necrotic tumor and radiation-related side effects.

In conclusion, SRS represents a therapeutic approach for recurrent or residual ependymomas, particularly when combined with standard radiation, chemotherapy, and maximum safe surgical resection.

5.12   CRANIOPHARYNGIOMAS

Craniopharyngiomas, benign neuroepithelial tumors originating from remnants of Rathke's pouch or the hypophyseal duct, exhibit a slow growth rate. Given their proximity to vital structures like the pituitary stalk, optic apparatus, hypothalamus, and blood vessels, a customized multimodal treatment approach is essential for each patient. GTR is the preferred option whenever feasible. STR is utilized for diagnostic confirmation, tumor debulking, and alleviating mass effect when complete removal is not achievable. Often, radiation therapy accompanies surgical interventions.

Research on CyberKnife radiosurgery for craniopharyngiomas is limited compared to Gamma Knife radiosurgery. In a small trial involving 11 participants, researchers reported a 91% tumor control rate after a mean follow-up of 15.4 months [62]. Another study with 43 individuals revealed local control rates of 85% at three years and 65% at five years [63]. GK radiosurgery is most effective for smaller, predominantly solid tumors located ideally a few millimeters away from the optic system. However, it can also treat smaller tumors (<3 cm in diameter) of any type.

5.13   CONCLUSION

SRS represents a significant advancement in the treatment of brain tumors, offering targeted therapy with minimal damage to surrounding healthy tissues. This chapter has outlined the principles, applications, and clinical efficacy of SRS, demonstrating its superiority in managing both benign and malignant brain tumors. The ability of SRS to adapt to the complex anatomy and sensitivity of brain structures while maintaining effective tumor control is underscored by its integration of advanced radiobiological principles and precision targeting techniques. Future research is required to further refine SRS protocols and explore its combination with other therapies to enhance outcomes, particularly in the context of tumor resistance and recurrence. The evolving landscape of SRS in neuro-oncology promises continued improvements in patient survival rates and quality of life, making it an indispensable tool in the modern therapeutic arsenal against brain tumors.

REFERENCES

[1] Valiente, M. et al. The evolving landscape of brain metastasis. Trends Cancer 4, 176–196 (2018).
[2] Boire, A. et al. Complement component 3 adapts the cerebrospinal fluid for leptomeningeal metastasis. Cell 168, 1101–1113 (2017).
[3] Neman, J. et al. Human breast cancer metastases to the brain display GABAergic properties in the neural niche. Proc. Natl Acad. Sci. USA 111, 984–989 (2014).
[4] Berghoff, A. S. et al. Density of tumor-infiltrating lymphocytes correlates with extent of brain edema and overall survival time in patients with brain metastases. Oncoimmunology 5, e1057388 (2016).
[5] Sevenich, L. et al. Analysis of tumour- and stroma-supplied proteolytic networks reveals a brain-metastasis-promoting role for cathepsin S. Nat. Cell Biol. 16, 876–888 (2014).
[6] Tawbi, H. A. et al. Combined nivolumab and ipilimumab in melanoma metastatic to the brain. N. Engl. J. Med. 379, 722–730 (2018).
[7] Long, G. V. et al. Combination nivolumab and ipilimumab or nivolumab alone in melanoma brain metastases: A multicentre randomised phase 2 study. Lancet Oncol. 19, 672–681 (2018).
[8] Lun, M. P. et al. Development and functions of the choroid plexus–cerebrospinal fluid system. Nat. Rev. Neurosci. 16, 445–457 (2015).
[9] Valiente, M. et al. Serpins promote cancer cell survival and vascular co-option in brain metastasis. Cell 156, 1002–1016 (2014).
[10] Zhang, L. et al. Microenvironment-induced PTEN loss by exosomal microRNA primes brain metastasis outgrowth. Nature 527, 100–104 (2015).
[11] Priego, N. et al. STAT3 labels a subpopulation of reactive astrocytes required for brain metastasis article. Nat. Med. 24, 1481 (2018).
[12] da Fonseca, A. C. C. et al. The impact of microglial activation on blood–brain barrier in brain diseases. Front. Cell. Neurosci. 8, 362 (2014).
[13] Seano, G. et al. Solid stress in brain tumours causes neuronal loss and neurological dysfunction and can be reversed by lithium. Nat. Biomed. Eng. 3, 230–245 (2019).
[14] Venkatesh, H. S. et al. Neuronal activity promotes glioma growth through neuroligin-3 secretion. Cell 161, 803–816 (2015).
[15] Venkatesh, H. S. et al. Electrical and synaptic integration of glioma into neural circuits. Nature 573, 539–545 (2019).

[16] Zeng, Q. et al. Synaptic proximity enables NMDAR signalling to promote brain metastasis. Nature 573, 526–531 (2019).

[17] Yoo, B. C. et al. Cerebrospinal fluid metabolomic profiles can discriminate patients with leptomeningeal carcinomatosis from patients at high risk for leptomeningeal metastasis. Oncotarget 8, 101203–101214 (2017).

[18] Gholamin, S. et al. Disrupting the CD47–SIRPα anti-phagocytic axis by a humanized anti-CD47 antibody is an efficacious treatment for malignant pediatric brain tumors. Sci. Transl Med. 9, eaaf2968 (2017).

[19] Zhang, X. H.-F. et al. Selection of bone metastasis seeds by mesenchymal signals in the primary tumor stroma. Cell 154, 1060–1073 (2013).

[20] Louie, E. et al. Neurotrophin-3 modulates breast cancer cells and the microenvironment to promote the growth of breast cancer brain metastasis. Oncogene 32, 4064–4077 (2013).

[21] Chen, Q. et al. Carcinoma–astrocyte gap junctions promote brain metastasis by cGAMP transfer. Nature 533, 493–498 (2016).

[22] Venkataramani, V. et al. Glutamatergic synaptic input to glioma cells drives brain tumour progression. Nature 573, 532–538 (2019).

[23] Lockman, P. R. et al. Heterogeneous blood–tumor barrier permeability determines drug efficacy in experimental brain metastases of breast cancer. Clin. Cancer Res. 16, 5664–5678 (2010).

[24] Lyle, L. T. et al. Alterations in pericyte subpopulations are associated with elevated blood–tumor barrier permeability in experimental brain metastasis of breast cancer. Clin. Cancer Res. 22, 5287–5299 (2016).

[25] Griveau, A. et al. A glial signature and Wnt7 signaling regulate glioma–vascular interactions and tumor microenvironment. Cancer Cell 33, 874–889.e7 (2018).

[26] Brastianos, P. K. et al. Exome sequencing identifies BRAF mutations in papillary craniopharyngiomas. Nat. Genet. 46, 161–165 (2014).

[27] Juratli, T. A. et al. Targeted treatment of papillary craniopharyngiomas harboring BRAF V600E mutations. Cancer 125, 2910–2914 (2019).

[28] Drilon, A. et al. Safety and antitumor activity of the multitargeted Pan-TRK, ROS1, and ALK inhibitor entrectinib: Combined results from two phase I trials (ALKA-372–001 and STARTRK-1). Cancer Discov. 7, 400–409 (2017).

[29] Jacob, L. S. et al. Metastatic competence can emerge with selection of preexisting oncogenic alleles without a need of new mutations. Cancer Res. 75, 3713–3719 (2015).

[30] Greaves, M. & Maley, C. C. Clonal evolution in cancer. Nature 481, 306–313 (2012).

[31] Shah, S. P. et al. The clonal and mutational evolution spectrum of primary triple-negative breast cancers. Nature 486, 395–399 (2012).

[32] Bowman, R. L. et al. Macrophage ontogeny underlies differences in tumor-specific education in brain malignancies. Cell Rep. 17, 2445–2459 (2016).

[33] Seute, T. et al. Response of asymptomatic brain metastases from small-cell lung cancer to systemic first-line chemotherapy. J. Clin. Oncol. 24, 2079–2083 (2006).

[34] Qian, B. Z. et al. CCL2 recruits inflammatory monocytes to facilitate breast–tumour metastasis. Nature 475, 222–475 (2011).

[35] Garzia, L. et al. A hematogenous route for medulloblastoma leptomeningeal metastases. Cell 172, 1050–1062.e14 (2018).

[36] Taggart, D. et al. Anti-PD-1/anti-CTLA-4 efficacy in melanoma brain metastases depends on extracranial disease and augmentation of CD8+ T cell trafficking. Proc. Natl Acad. Sci. USA 115, E1540–E1549 (2018).

[37] Cao, K. I. et al. Phase II randomized study of whole-brain radiation therapy with or without concurrent temozolomide for brain metastases from breast cancer. Ann. Oncol. 26, 89–94 (2015).

[38] Palmieri, D. et al. Profound prevention of experimental brain metastases of breast cancer by temozolomide in an MGMT-dependent manner. Clin. Cancer Res. 20, 2727–2739 (2014).

[39] Sunwoo, L. et al. Differentiation of glioblastoma from brain metastasis: Qualitative and quantitative analysis using arterial spin labeling MR imaging. PLoS One 11, e0166662 (2016).

[40] Kuczynski, E. A. et al. Vessel co-option in cancer. Nat. Rev. Clin. Oncol. 16, 469–493 (2019).

[41] Mashimo, T. et al. Acetate is a bioenergetic substrate for human glioblastoma and brain metastases. Cell 159, 1603–1614 (2014).

[42] Bos, P. D. et al. Genes that mediate breast cancer metastasis to the brain. Nature 459, 1005–1009 (2009).

[43] Ippen, F. M. et al. The dual PI3K/mTOR pathway inhibitor GDC-0084 achieves antitumor activity in PIK3CA-mutant breast cancer brain metastases. Clin. Cancer Res. 25, 3374–3383 (2019).

[44] Meuwissen, R. et al. Induction of small cell lung cancer by somatic inactivation of both Trp53 and Rb1 in a conditional mouse model. Cancer Cell 4, 181–189 (2003).

[45] Kato, M. et al. Transgenic mouse model for skin malignant melanoma. Oncogene 17, 1885–1888 (1998).

[46] Cho, J. H. et al. AKT1 activation promotes development of melanoma metastases. Cell Rep. 13, 898–905 (2015).

[47] Wu, X. et al. Clonal selection drives genetic divergence of metastatic medulloblastoma. Nature 482, 529–533 (2012).

[48] Moriarity, B. S. et al. A sleeping beauty forward genetic screen identifies new genes and pathways driving osteosarcoma development and metastasis. Nat. Genet. 47, 615–624 (2015).

[49] De La Rochere, P. et al. Humanized mice for the study of immuno-oncology. Trends Immunol. 39, 748–763 (2018).

[50] Winslow, M. M. et al. Suppression of lung adenocarcinoma progression by Nkx2–1. Nature 473, 101–104 (2011).

[51] Schade, B. et al. PTEN deficiency in a luminal ErbB-2 mouse model results in dramatic acceleration of mammary tumorigenesis and metastasis. J. Biol. Chem. 284, 19018–19026 (2009).

[52] Kircher, D. A. et al. AKT1E17K activates focal adhesion kinase and promotes melanoma brain metastasis. Mol. Cancer Res. 17, 1787–1800 (2019).

[53] Fischer, G. M. et al. Molecular profiling reveals unique immune and metabolic features of melanoma brain metastases. Cancer Discov. 9, 628–645 (2019).

[54] Louveau, A. et al. Structural and functional features of central nervous system lymphatic vessels. Nature 523, 337–341 (2015).

[55] Louveau, A. et al. CNS lymphatic drainage and neuroinflammation are regulated by meningeal lymphatic vasculature. Nat. Neurosci. 21, 1380–1391 (2018).

[56] Iliff, J. J. et al. A paravascular pathway facilitates CSF flow through the brain parenchyma and the clearance of interstitial solutes, including amyloid β. Sci. Transl Med. 4, 147ra111 (2012).

[57] Lin, N. U. et al. Response assessment criteria for brain metastases: Proposal from the RANO group. Lancet Oncol. 16, e270–e278 (2015).

[58] Timmer, M. et al. Discordance and conversion rates of progesterone-, estrogen-, and HER2/neu-receptor status in primary breast cancer and brain metastasis mainly triggered by hormone therapy. Anticancer Res. 37, 4859–4865 (2017).

[59] Priedigkeit, N. et al. Intrinsic subtype switching and acquired ERBB2/HER2 amplifications and mutations in breast cancer brain metastases. JAMA Oncol. 3, 666–671 (2017).

[60] Brastianos, P. K. et al. Genomic characterization of brain metastases reveals branched evolution and potential therapeutic targets. Cancer Discov. 5, 1164–1177 (2015).

[61] Nygaard, V. et al. Melanoma brain colonization involves the emergence of a brain-adaptive phenotype. Oncoscience 1, 82–94 (2014).

[62] Park, E. S. et al. Cross-species hybridization of microarrays for studying tumor transcriptome of brain metastasis. Proc. Natl Acad. Sci. USA 108, 17456–17461 (2011).

[63] Posner, J. B. Management of brain metastases. Rev. Neurol. 148, 477–487 (1992).

6 Antisense Therapy in Combating Brain Cancer

Xiaodong Li[1]*, Sayantani Garai[2], Dipro Mukherjee[2], and Nivedita Chatterjee[2]

[1]Department of Neurosurgery, Shengjing Hospital of China Medical University, Shenyang, Liaoning, China
[2]Department of Biotechnology, University of Engineering & Management, Kolkata West Bengal, India
*Corresponding author

6.1 INTRODUCTION

Presently, most drugs target specific protein molecules, receptors, enzymes, or other downstream molecules involved in the molecular basis of diseases [1]. This scenario is mirrored in brain tumors. However, antisense technology diverges by targeting RNAs instead of downstream products. This technology employs small oligonucleotide sequences as promising therapeutics to suppress disease-causing factors at a genetic level. These oligonucleotides are designed to complement specific RNA sequences encoded by target genes. Upon administration, they bind to these RNA sequences through Watson–Crick base pairing, modulating the function of the target RNA [1].

Targeting RNA instead of downstream products offers several advantages as a therapeutic strategy, significantly broadening the range of potential targets for therapy. In brain cancer, antisense oligonucleotides (ASOs) are known to inhibit the expression of many cancer-specific genes by inhibiting transcription or translation [2]. In contrast to conventional treatments for brain cancer, such as radiation therapy and chemotherapy, which lack specificity and often harm adjacent normal cells, antisense therapy is highly specific, targeting only mutated or cancer-associated genetic products [3].

The development of antisense therapeutics has become a clinical reality and holds immense promise in brain cancer therapy [3].

6.2 PHARMACOLOGY OF ANTISENSE DRUGS

Antisense drugs, typically consisting of short oligonucleotide sequences (12–24-mers) [4], act by binding to RNAs encoded by target genes through basic Watson–Crick complementary base pairing. This interaction leads to modulations that either enhance or suppress RNA activity. Such modulations can occur through various mechanisms, including binding to RNA and interfering with its function without degrading the target. This can be achieved by modulating RNA processing, displacing bound proteins, or inducing translation arrest.

Another mechanism by which ASOs exert their effects is by promoting RNA degradation through RNA interference (RNAi) [5]. Some literature suggests that ASOs can enhance or regulate protein production by correcting faulty RNA splicing or by antagonizing microRNAs (miRNAs) that suppress protein production [6].

To overcome the inherent instability of DNA and RNA molecules in biological systems, various chemical modifications and formulations are employed to protect them from enzymatic and environmental degradation. One widely adopted strategy involves the chemical modification of oligonucleotides, which increases their stability in biological environments by shielding them from

DOI: 10.1201/9781003519706-6

endogenous nucleases. Additionally, these modifications enhance therapeutic efficiency by improving the oligonucleotides' affinity toward target RNA molecules.

For instance, chemical modification of the phosphodiester linkages between the monomers of oligonucleotides using methyl phosphonate, phosphothiolate, and phosphorodiamidate-based chemistries has been performed. Currently, many antisense drugs utilize phosphothiorate modification, which not only enhances stability but also improves tissue distribution and cellular uptake of the drugs [4]. Furthermore, heterocycle and sugar modifications are known to enhance the drug-like properties of oligonucleotides.

6.3   MECHANISMS OF ACTION

6.3.1   IGF-I ANTISENSE AND TRIPLE-HELIX GENE THERAPY

For cytotoxic CD8+ T cells to exert their effect, a bridge must form between CD8 and class I major histocompatibility complex (MHC-I) antigens. Transfection with antisense cDNA of insulin-like growth factor I (IGF-I) has the potential to lead to a fivefold increase in MHC-I expression in glioma cells and in vivo. While this increase in MHC-I expression may not be the sole mechanism involved in the cytotoxic response, it holds the potential to play a significant role [7].

In rats with tumors, the injection of transfected cells resulted in an exceptionally strong antitumor immune response, leading to a high yield of CD8+ cells. This robust immune response effectively halted the progression of the cells into further tumor development [8]. Treatment involving the use of anti-IGF-I cDNA in an antisense orientation showed promising results. In vitro transfection of tumor cells with cDNA containing IGF-I antisense RNA resulted in the production of IGF-I antisense RNA, which was negatively labeled with anti-IGF-I antibodies and positively labeled with anti-MHC-I and anti-B7 antibodies [9]. However, approximately half of the transfected cells underwent apoptosis, highlighting the interconnectedness of apoptosis and immune reactions through signal transduction pathways [10]. Transfected cells, including apoptotic ones, in combination with induced in vivo antigen-presenting cells, are injected into the patient, activating T lymphocytes (CTL CD8+ CD28+). These activated T lymphocytes can mount a potent anti-tumor immune response [11].

Immunological responses to treated tumors, mediated by tumor antigen-specific CD4+ and CD8+ T cells, appear to be a primary mechanism of tumor destruction induced by antisense [12] (Figure 6.1).

6.3.2   ASO DRUG DELIVERY

Due to their duality, ASOs have two distinct mechanisms of action [13–15]. The first involves the degradation of RNA, while the second involves the inhibition or modulation of RNA through steric hindrance [16, 17] (Figure 6.2).

6.4   STRATEGIES OF ASO DRUG DELIVERY TO BRAIN

Delivery of ASO drugs to the site of action presents one of the biggest challenges, involving their injection into the bloodstream, traversal of biological barriers, and internalization by cells [18]. Once internalized, ASOs must evade degradation by lysosomes and avoid entrapment in secretory vesicles that transport cargo from cells to the surface [19]. To enhance a drug's stability, targeting, and trafficking, various strategies have been extensively researched, including chemical modifications, bioconjugation to different moieties, and the use of delivery mechanisms.

The first chemical modification, marking the inception of the three generations of ASOs, involves modifying the internucleotide phosphate group [20]. This modification entails adding a sulfur atom to a non-bridging oxygen atom, thus creating phosphorothioate (PS) backbones. PS–ASOs exhibit

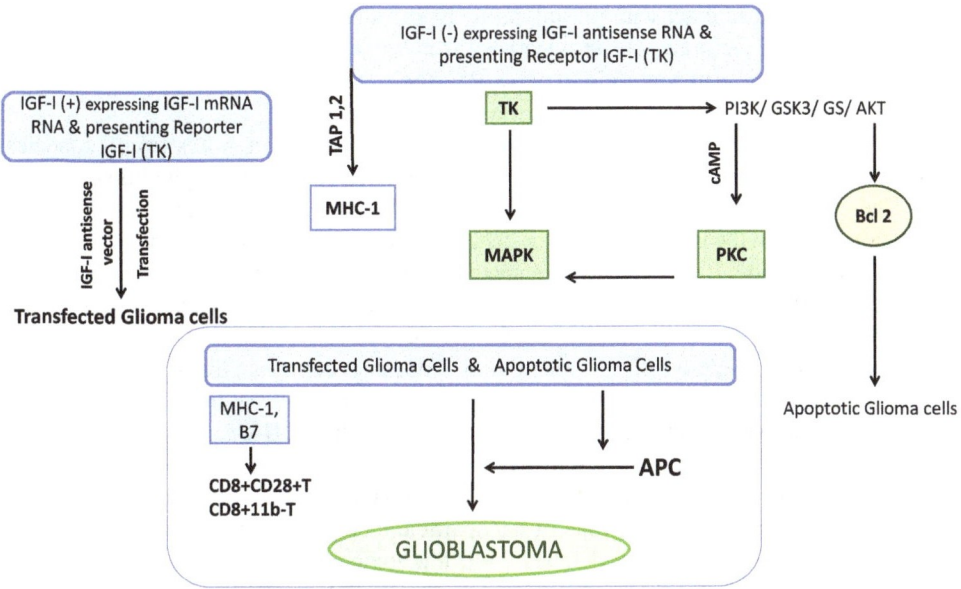

FIGURE 6.1 Mechanism of action for IGF-I antisense therapy in glioblastoma. The diagram illustrates the process of transfecting glioma cells with an IGF-I antisense vector, leading to the expression of IGF-I antisense RNA and IGF-I receptor presentation. This triggers a cascade involving MHC-1, MAPK, PKC, and PI3K pathways, ultimately leading to apoptosis in glioma cells. The lower section of the diagram shows the immune response involving CD8+ T cells and antigen-presenting cells that target glioblastoma cells.

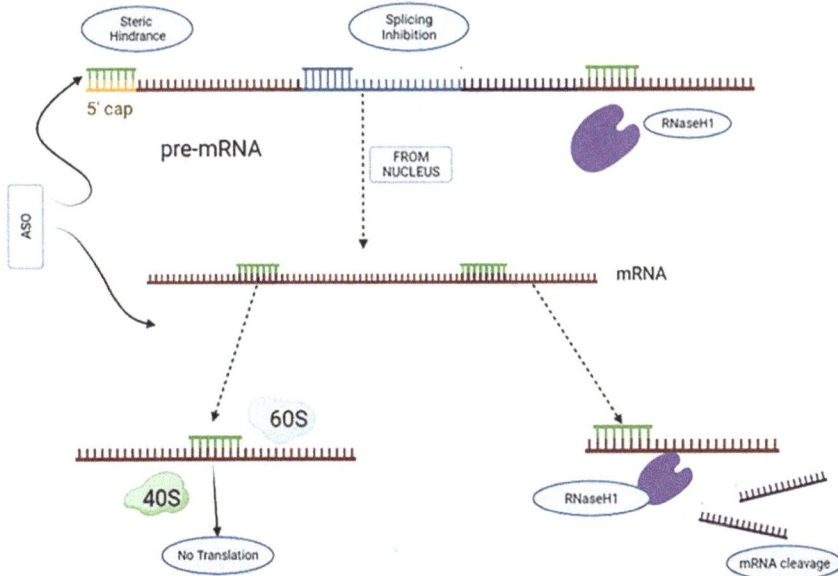

FIGURE 6.2 Mechanisms of antisense oligonucleotides (ASOs) in modulating pre-mRNA and mRNA. The diagram illustrates how ASOs can bind to pre-mRNA to cause steric hindrance, splicing inhibition, or RNaseH1-mediated degradation. ASOs can also bind to mRNA, leading to either translation inhibition by preventing ribosomal assembly or mRNA cleavage through RNaseH1 activity.

nuclease resistance and interact with plasma proteins like albumin, resulting in a longer half-life. This extended half-life facilitates slower clearance from the body and enables binding to proteins throughout the endocytic pathway. Additionally, PS–ASOs interact with various intracellular proteins, which can either enhance or impede their activity within the cell [21].

Once the necessary chemical modifications are identified to protect ASOs from exonucleases and extend their stability, the next challenge is to facilitate their passage through biological barriers [21, 22]. These barriers include the vascular endothelial barrier, cell membrane, and intracellular compartments. Thus far, the liver, eyes, and the central nervous system have been the most accessible targets for delivery, leading to a focus on therapeutics for these areas (e.g., Inotersen, Formivirsen, Nusinersen) [22–29]. ASO drugs tend to distribute and accumulate in the eye and spinal cord due to the gaps present along the endothelial linings of the liver.

One strategy to enhance the delivery potential of ASOs is to conjugate them with different moieties, directing them to specific tissues and promoting internalization [23]. In recent years, a variety of nano-drug delivery vehicles have emerged in studies, including DNA nanostructures, exosome-like nanocarriers, and spherical nucleic acids, as nanotechnology rapidly advances [24–26].

Liposomes, composed of natural or synthetic lipid bilayers surrounding an internal aqueous compartment, have become a cornerstone of nanoparticle therapeutics since their discovery in the 1960s. They offer stability, selective delivery [30], low toxicity, and biocompatibility, which are significant advantages [31–35]. Currently, numerous liposome formulations are undergoing clinical trials. These liposomes, with varying charges, sizes, and efficacy, are developed by combining various components such as lipids, synthetic amphiphiles, and sterols [36, 37]. Some of the most recent advances in liposomal technology described for optimizing drug delivery are summarized in Figure 6.3 [38, 39].

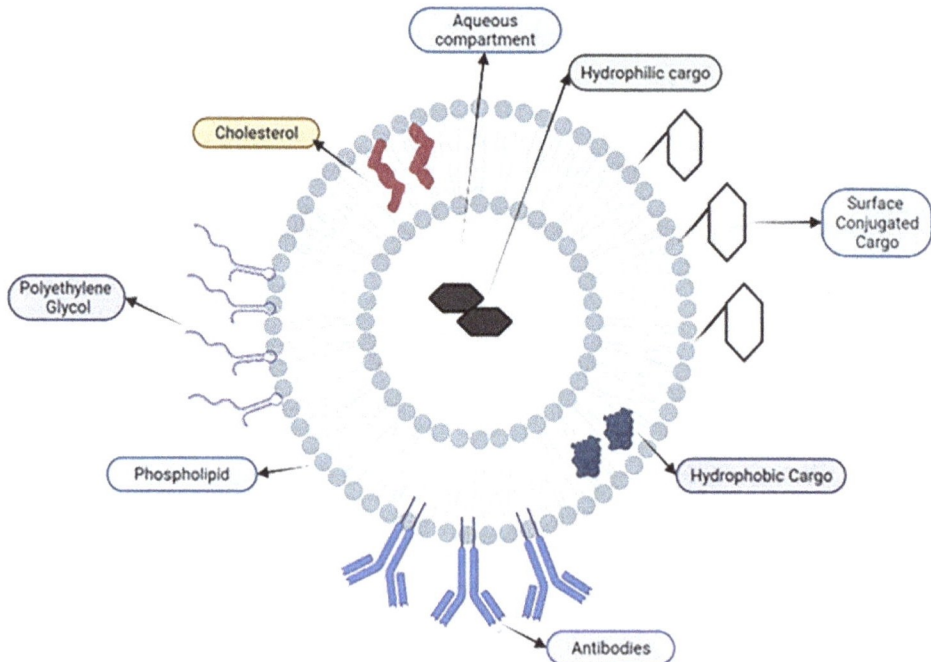

FIGURE 6.3 Schematic representation of a liposome used for drug delivery. The diagram shows the structure of a liposome, including the aqueous compartment for hydrophilic cargo, the phospholipid bilayer for hydrophobic cargo, and surface-conjugated cargo. Key components include polyethylene glycol (PEG) for stability, cholesterol for membrane rigidity, phospholipids for the bilayer structure, and antibodies for targeting.

6.5 ANTISENSE THERAPY FOR BRAIN CANCER: PAST AND PROSPECTS

ASOs bind to RNA, altering its function and paving the way for the development of antisense-based drugs and medicine. From a therapeutic standpoint, this approach influences RNA processing and modulates the expression of challenging-to-target proteins encoded by the bound RNAs [40]. While the translation of this technology into clinical practice encountered drawbacks such as target arrangement, biological activity, and off-target effects over the past two decades [41], it has significantly broadened the range and types of targets available for treatment, thereby enhancing their effectiveness. Moreover, this technology offers a notable advantage in developing drugs for both monogenic and polygenic diseases and in influencing human genetics and genomics [41, 42]. By reducing unwanted and harmful proteins in the body, it holds the potential to halt the progression of certain degenerative diseases.

Over the years, novel chemical modifications have been employed to address the limitations associated with ASOs [41]. These modifications, along with advancements in clinical trial design and understanding of antisense alterations and their mechanisms of action, have propelled the translation of ASO-based strategies into therapies [43].

The clinical application of antisense pharmacology in humans has been convincingly demonstrated through multiple clinical trials, showcasing their safety across various mechanisms. Currently, antisense drugs are undergoing development in clinical trials, targeting a wide array of tissues both locally and systemically [42]. Leveraging advanced oligonucleotide chemistry further enhances the properties of antisense drugs, increasing potency and safety and enabling broader tissue distribution.

Antisense pharmacology has demonstrated clinical applications in humans, with safety established through various mechanisms in multiple clinical trials. Current antisense drugs in clinical development target a diverse range of tissues systemically and locally [43]. Leveraging advanced oligonucleotide chemistry has further enhanced antisense drugs, increasing their potency, safety, and broad tissue distribution. ASOs have shown significant potential for genomic modification, including selective gene silencing at the transcription or translation level, as well as targeted gene expression suppression.

Antisense pharmacology has proven its clinical relevance in human applications, with safety confirmed through various mechanisms in numerous clinical trials. Presently, antisense drugs in clinical development are designed to target a wide spectrum of tissues both systemically and locally [43]. The utilization of advanced oligonucleotide chemistry has notably improved antisense drugs, amplifying their effectiveness, safety, and ability to reach diverse tissue types. Moreover, ASOs exhibit promising capabilities for genomic modification, encompassing selective gene silencing at either the transcription or translation level, as well as targeted suppression of gene expression.

Initially, there has always been a delay between the discovery of a therapy and its testing in clinical trials and subsequent release to the market. RNA-based precision medicine and gene therapy, due to their off-target effects, still require improvement. The future applications of antisense technology will only be discerned through the evaluation of novel molecules in clinical trials [42, 43].

Significant strides have been made in antisense technology, yet its full potential remains unrealized. Numerous unanswered questions persist, necessitating further advancements in this field. The advent of first-generation PS oligodeoxynucleotides has presented patients with novel treatment options, proving to be a valuable asset and pharmacological tool [44].

In the future, advancements in ASOs, featuring innovative formulations, hold promise for enhanced patient treatments. Despite significant progress in recent years, unanswered questions persist in antisense technology, leaving room for further refinement. Although much has been achieved, opportunities for improvement on this platform still abound [40].

Additionally, a significant advancement in RNA therapeutics is the emergence of CRISPR, a gene editing technology guided by RNA, alongside the use of in vitro transcribed mRNA as a treatment option [45]. Both in laboratory settings and clinical trials, RNA-based antisense drugs are undergoing testing for their ability to modulate gene/protein expression and genome editing. The

CRISPR-Cas genome editing technology has revolutionized genome editing, profoundly impacting the biomedical science sector and facilitating the development of antisense RNA-based delivery approaches for clinical translation [44, 45]. Moreover, the development of nanomedicine holds promise in overcoming the limitations of conventional oligonucleotide delivery in vivo by enabling controlled and sustained drug release.

6.6 CONCLUSION

An ASO is a short, synthetic, single-stranded oligodeoxynucleotide that targets RNA to reduce, restore, or modify protein expression at the source of pathogenesis, making it more effective than downstream therapies. Recent advancements in understanding antisense pharmacology have increased confidence in translating antisense therapeutics into clinical settings. However, further optimization of antisense delivery, target engagement, and safety profiles is necessary to advance this technology in clinical practice. Antisense technology holds the promise of transforming the treatment of several disease conditions in the near future.

Recently, the FDA approved several treatments based on gene therapy, including Gendicine (Ad-p53) for squamous cell carcinoma, Vitravene (fomivirsen) for cytomegalovirus retinitis, Macugen (pegaptanib) for age-related macular degeneration, Kynamro (mipomersen) for homozygous familial hypercholesterolemia, Exondys 51 (eteplirsen) for Duchenne muscular dystrophy, Defitelio (defibrotide) for severe hepatic veno-occlusive disease, and Spinraza (nusinersen) for spinal muscular atrophy. Advancements in design, chemistry, synthesis, and delivery technologies have led to the improved stability, efficacy, specificity, and immune evasion of antisense technology, showcasing its growing potential.

REFERENCES

[1] Bennett CF, Kordasiewicz HB, Cleveland DW. Antisense drugs make sense for neurological diseases. Annual Review of Pharmacology and Toxicology 2021; 61(1): 831–852. DOI: 10.1146/annurev-pharmtox-010919-023738

[2] Eckstein F. Phosphorothioate oligodeoxynucleotides: What is their origin and what is unique about them? Antisense and Nucleic Acid Drug Development 2000; 10: 117–121.

[3] Caruso G, Caffo M. Antisense oligonucleotides in the treatment of cerebral gliomas. Review of concerning patents. Recent Patents on CNS Drug Discovery 2014; 9(1): 2–12. DOI: 10.2174/1574889809666614 0307113439

[4] Crooke ST, Baker BF, Crooke RM, Liang X. Antisense technology: An overview and prospectus. Nature Reviews Drug Discovery 2021; 20(6): 427–453. DOI: 10.1038/s41573-021-00162-z

[5] Bennett CF, Baker BF, Pham N, Swayze E, Geary RS. Pharmacology of antisense drugs. Annual Review of Pharmacology and Toxicology 2017; 57(1). DOI: 10.1146/annurev-pharmtox-010716-104846

[6] Liang X-H, Shen W, Sun H, Migawa MT, Vickers TA, Crooker ST. Translation efficiency of mRNAs is increased by antisense oligonucleotides targeting upstream open reading frames. Nature Biotechnology 2016; 34: 875–880.

[7] Brooks WH, Latta RB, Mahaley MS. Immunobiology of primary intracranial tumors. Journal of Neurosurgery 1981; 54: 331–337.

[8] Schwartz RH. Costimulation of T lymphocytes: The role ofCD28, CTLA-4 and B7/BBI in interleukin-2 production and immunotherapy. Cell 1992; 71: 1065–1068.

[9] Ly A, Duc HT, Kalamarides M, Trojan LA, Pan Y, Shevelev A, François J-C, Noël T, Kane A, Henin D, Anthony DD, Trojan J. Human glioma cells transformed by IGF-I triple-helix technology show immune and apoptotic characteristics determining cell selection for gene therapy of glioblastoma. Journal of Clinical Pathology (Molecular Pathology) 2001; 54(4): 230–239.

[10] Trojan J, Cloix J-F, Ardourel M-Y, Chatel M, Anthony D. IGF-I biology and targeting in malignant glioma. Neuroscience 2007; 145(3): 795–812.

[11] Trojan J, Ly A, Wei MX, Kopinski P, Ardourel M-Y, Pan Y, Trojan LA, Dufour D, Shevelev A, Andres C, Chatel M, Kasprzak H, Anthony DD, Duc HT. Antisense anti-IGF-I cellular therapy of malignant tumours: Immune response in cancer patients. Biomedicine & Pharmacotherapy 2010; 64(8): 576–578.

[12] Trojan J, Briceno I. IGF-1 Antisense and tripple-helix gene therapy of glioblastoma. 2013. DOI: 10.5772/52366

[13] Liang X-H, Sun H, Nichols JG, Crooke ST. RNase H1-dependent antisense oligonucleotides are robustly active in directing RNA cleavage in both the cytoplasm and the nucleus. Molecular Therapy 2017; 25: 2075–2092.

[14] Liang X-H, Nichols JG, De Hoyos CL, Crooke ST. Some ASOs that bind in the coding region of mRNAs and induce RNase H1 cleavage can cause increases in the pre-mRNAs that may blunt total activity. Nucleic Acids Research 2020; 48: 9840–9858.

[15] Dominski Z, Kole R. Restoration of correct splicing in thalassemic pre-mRNA by antisense oligonucleotides. Proceedings of the National Academy of Sciences of the United States of America 1993; 90: 8673–8677.

[16] Liang X-H, Shen W, Sun H, Migawa MT, Vickers TA, Crooke ST. Translation efficiency of mRNAs is increased by antisense oligonucleotides targeting upstream open reading frames. Nature Biotechnology 2016; 34: 875–880.

[17] Somers J, Pöyry T, Willis AE. A perspective on mammalian upstream open reading frame function. The International Journal of Biochemistry & Cell Biology 2013; 45: 1690–1700.

[18] Linnane E, Davey P, Zhang P, Puri S, Edbrooke M, Chiarparin E, Revenko AS, MacLeod AR, Norman JC, Ross SJ. Differential uptake, kinetics and mechanisms of intracellular trafficking of next-generation antisense oligonucleotides across human cancer cell lines. Nucleic Acids Research 2019; 47: 4375–4392.

[19] Thomas OS, Weber W. Overcoming physiological barriers to nanoparticle delivery—are we there yet? Frontiers in Bioengineering and Biotechnology 2019; 7: 415.

[20] Crooke ST, Vickers TA, Liang X-H. Phosphorothioate modified oligonucleotide–protein interactions. Nucleic Acids Research 2020; 48: 5235–5253.

[21] Gaus HJ, Gupta R, Chappell AE, Østergaard ME, Swayze EE, Seth PP. Characterization of the interactions of hemicallymodified therapeutic nucleic acids with plasma proteins using a fluorescence polarization assay. Nucleic Acids Research 2018; 47: 1110–1122.

[22] Liang X-H, Sun H, Shen W, Crooke ST. Identification and characterization of intracellular proteins that bind oligonucleotides with phosphorothioate linkages. Nucleic Acids Research 2015; 43: 2927–2945.

[23] Shen X, Corey DR. Chemistry, mechanism and clinical status of antisense oligonucleotides and duplex RNAs. Nucleic Acids Research 2018; 46: 1584–1600.

[24] Lundin KE, Gissberg O, Smith CE. Oligonucleotide therapies: The past and the present. Human Gene Therapy 2015; 26: 475–485.

[25] Copolovici DM, Langel K, Eriste E, Langel Ü. Cell-penetrating peptides: Design, synthesis, and applications. ACS Nano 2014; 8: 1972–1994.

[26] McClorey G, Banerjee S. Cell-penetrating peptides to enhance delivery of oligonucleotide-based therapeutics. Biomedicines 2018; 6: 51.

[27] Ashizawa AT, Cortes J. Liposomal delivery of nucleic acid-based anticancer therapeutics: BP-100–1.01. Expert Opinion on Drug Delivery 2014; 12: 1107–1120.

[28] Daraee H, Etemadi A, Kouhi M, Alimirzalu S, Akbarzadeh A. Application of liposomes in medicine and drug delivery. Artificial Cells, Nanomedicine, and Biotechnology 2016; 44: 381–391.

[29] Bozzuto G, Molinari A. Liposomes as nanomedical devices. International Journal of Nanomedicine 2015; 10: 975–999.

[30] Maclachlan I. Liposomal formulations for nucleic acid delivery. In Antisense Drug Technology; Crooke, S.T., Ed.; CRC Press: Boca Raton, FL, 2007; pp. 237–270.

[31] Abu Lila AS, Kiwada H, Ishida T. The accelerated blood clearance (ABC) phenomenon: Clinical challenge and approaches to manage. Journal of Controlled Release 2013; 172: 38–47.

[32] Zelphati O, Uyechi LS, Barron LG, Szoka FC. Effect of serum components on the physico-chemical properties of cationic lipid/oligonucleotide complexes and on their interactions with cells. Biochimica et Biophysica Acta (BBA) – Lipids and Lipid Metabolism 1998; 1390: 119–133.

[33] Nag OK, Awasthi V. Surface engineering of liposomes for stealth behavior. Pharmaceutics 2013; 5: 542–569.

[34] Gabizon AA. Pegylated liposomal doxorubicin: Metamorphosis of an old drug into a new form of chemotherapy. Cancer Investigation 2001; 19: 424–436.

[35] Schöttler S, Becker G, Winzen S, Steinbach T, Mohr K, Landfester K, Mailänder SSV, Wurm F. Protein adsorption is required for stealth effect of poly(ethylene glycol)- and poly(phosphoester) coated nanocarriers. Nature Nanotechnology 2016; 11: 372–377.

[36] Thi TTH, Pilkington EH, Nguyen DH, Lee JS, Park KD, Truong NP. The importance of poly(ethylene glycol) lternatives for overcoming PEG immunogenicity in drug delivery and bioconjugation. Polymers 2020; 12: 298.

[37] Romberg B, Oussoren C, Snel CJ, Carstens MG, Hennink WE, Storm G. Pharmacokinetics of poly(hydroxyethylasparagine)-coated liposomes is superior over that of PEG-coated liposomes at low lipid dose and upon repeated administration. Biochimica et Biophysica Acta (BBA) – Biomembranes 2007; 1768: 737–743.

[38] Barenholz Y. (Chezy) Doxil®—The first FDA-approved nano-drug: Lessons learned. Journal of Controlled Release 2012; 160: 117–134.

[39] Bennett CF, Baker BF, Pham N, Swayze E, Geary RS. Pharmacology of antisense drugs. Annual Review of Pharmacology and Toxicology 2017; 57: 81–105.

[40] Lundin KE, Gissberg O, Smith CI. Oligonucleotide therapies: The past and the present. Human Gene Therapy 2015; 26(8): 475–485.

[41] Krishna M. From petunias to the present—a review of oligonucleotide therapy. Journal of Clinical Epigenetics 2017; 3(4): 38.

[42] Geary RS, Norris D, Yu R, Bennett CF. Pharmacokinetics, biodistribution and cell uptake of antisense oligonucleotides. Advanced Drug Delivery Reviews 2015; 87: 46–51.

[43] Gulam M, Khan NJ, Maheshwari KK, Porwal M. Antisense therapy and its application in biopharmaceutical research: Past, present and future. Asian Pacific Journal of Applied Sciences 2016: 61–69.

[44] Kaczmarek JC, Kowalski PS, Anderson DG. Advances in the delivery of RNA therapeutics: From concept to clinical reality. Genome Medicine 2017; 9(1): 60.

[45] Sharad S. Antisense therapy: An overview. 2019. DOI: 10.5772/intechopen.86867.

7 Advanced Nanomaterial Applications in Brain Tumor Diagnosis and Theranostics

Yawen Ma[1]*, Rekha Khandia[2], and Pankaj Gurjar[3]
[1]Department of Neurosurgery, Shengjing Hospital of China Medical University, Shenyang, Liaoning, China
[2]Department of Biochemistry and Genetics, Barkatullah University, Bhopal MP, India
[3]Department of Science and Engineering, Novel Global Community Educational Foundation, Hebersham, Australia
*Corresponding author

7.1 INTRODUCTION

The brain is the most vital organ in the body to coordinate memory and motor functions. Aberrant growth of the cell in the brain region might disturb the normal functioning of the cell. The abnormal growth may be benign or malignant, and the tumors result in increased intracranial pressure, headaches, vomiting, loss of balance, altered consciousness, and seizures. Brain glial cells are common sites for a brain tumor. Gliomas are the most common kind of brain cancer; around 53.5% of all brain tumors are essentially gliomas. Brain tumors may be classified into four types (I to IV) based on their aggressiveness. Types III and IV are high-grade tumors, and more aggressive while I and II are of low grade [1] and less aggressive. Lung cancer in males [2] while breast cancer in females [3] metastasize to the brain. Various diagnosis tests for brain tumors include computed axial tomography (CAT scan/CT), magnetic resonance imaging (MRI), and positron emission tomography (PET scan). Diagnosis of brain tumors is still challenging owing to the non-specificity and heterogenous nature [4]. An improved detection system is needed to improve timely diagnosis with the least side effects compared to conventional diagnosis methodologies.

7.2 NANOPARTICLES IN MRI OF BRAIN TUMORS

MRI lacks contrast while imaging brain tumors where normal and tumor tissues are marginal. Thus to enhance the contrast, exogenous contrast agents (CAs) are added [5]. MRI enhancement is usually done using gadolinium-based contrast agents (GBCAs) [6]. However, it may lead to nephrogenic system fibrosis [7] and retention in the nervous system [8]; thus, alternative CAs are required. As an alternative, several iron oxide nanoparticles have been developed for a contrasting agent in MRI. In the work of Chan et al. [9], nanoparticles were combined with the MRI CAs. DOX and Fe–Pt nanoconjugate is injectable and may accumulate in the brain through magnetic induction. The Fe–Pt nanoconjugate displays high ferromagnetism and offers a promising imaging tool for detecting brain tumors (ferromagnetism). Nephrotoxicity and non-specificity limit the extensive use of Gd chelates in tumor MRI. Ultrasmall gadolinium oxide nanoparticles stabilized with poly(acrylic acid) (PAA) have been synthesized (ES-GON-PAA) and displayed a powerful T1-weighted MRI contrast. The specificity of tumor imaging with this nanoparticle may be further improved by adding an RGD dimer to target RGD overexpressing tumors. Nanocomposite is highly useful in magnetic tumor imaging owing to superior biocompatibility. Also, these are accumulated in the tumor. In the work of Duan et al. [10], nano agents were prepared for imaging and photothermal therapy (PTT). Cyclic peptide Arg-Gly-Asp (cRGD)-decorated ultrasmall iron oxide nanoparticles (UIONPs) of 5 nm size were nano-precipitated to form uniform spherical FULL nanocomposites with the addition of an amphiphilic polymer 1,2-distearoyl-sn-glycero-3-phosphoethanolamine-N-[methoxy (poly(ethylene glycol))-2000] (mPEG2000-DSPE) called FULL were prepared. In HALF, only at one surface were UIONPs present. cRGD ligand-attached nanocomposites showed higher cellular uptake by C6 cells in animal brain tumor models. Intravenous administration of nanocomposite enhanced signal intensity from 2 to 24 hours for MRI imaging.

7.3   NANOPARTICLES IN FLUOROMETRIC DETECTION

Flap endonuclease 1 (FEN1) is an enzyme found in elevated levels in various cancers, including glioma [11]. Its primary function involves removing Okazaki primers formed during DNA replication by cleaving single-stranded DNA (ssDNA) at the 5′ end of the replication fork. Cancer cells have heightened demands for rapid replication, leading to increased FEN1 levels, which correlate with cancer aggressiveness [12]. Moving away from conventional methods like ELISA, Western blotting, and immunohistochemistry, detection techniques are transitioning toward nanosensor-based approaches.

A dumbbell-shaped nanoprobe has been developed for this purpose, featuring a 5′ flap, a DNA–silver nanocluster (DNA–AgNCs), and a guanine-rich enhancer sequence in a 3′ flap. This probe's unique sequence and structure maintain its dumbbell shape, resulting in strong fluorescence. In the presence of FEN1, the enzyme cleaves the 5′ flap due to its specific activity, leading to reduced fluorescence. The degree of fluorescence reduction corresponds to the concentration of FEN1 in the sample, enabling sensitive detection with a limit of detection of 40 femtomolar [13]. This innovative sensor holds promise for widespread use in clinical settings for the early detection of brain cancers.

During surgery, distinguishing between brain tumor cells and healthy tissue can be challenging. To address this issue, researchers have developed triple-modality imaging nanoparticles, which combine magnetic resonance, photoacoustic, and Raman imaging capabilities. These nanoparticles are designed to penetrate the blood–brain barrier (BBB) only at sites where it is compromised, such as in areas of brain tumors. As a result, they accumulate specifically within the tumor tissue, making them valuable for guiding surgeons in delineating the boundaries of the tumor using any of the three imaging techniques mentioned [14].

Carbon dots synthesized from D-glucose and L-aspartic acid (CD-Asp) via the pyrolysis method exhibit remarkable targeting capabilities against glioma, a type of brain cancer. Interestingly, these carbon dots are able to detect glioma without the need for additional targeting agents. Moreover, they demonstrate excellent biocompatibility and the ability to emit full-color light, which can be adjusted as needed.

In animal studies, these carbon dots showed rapid biodistribution within just 15 minutes of injection via the tail vein, with notably strong signals detected specifically in the glioma-bearing brain compared to the control group. This dual functionality of CD-Asp provides both targetability and fluorescence imaging simultaneously, enabling non-invasive detection of glioma. This innovation represents a promising advancement in nanomedicine, where the integration of targeting, imaging, and therapy holds great potential [15].

Aptamers, small oligonucleotides known for their high specificity in binding to target molecules, can be obtained through a process called Systematic Evolution of Ligands via Exponential Enrichment (SELEX) [16]. Quantum dots are fluorescent nanocrystals with adjustable size, stability, and excellent biocompatibility [17–20]. Combining aptamers with quantum dots presents a promising approach for detecting brain tumor cells. One example involves the GBI-10 aptamer, which specifically targets tenascin-C, a protein overexpressed on the surface of glioma cells. By conjugating GBI-10 aptamers to quantum dots, a potent detection system can be developed for efficiently and conveniently identifying brain tumor cells [20].

7.4 CAMOUFLAGING NANOPARTICLES

Recently, there has been growing interest in cell membrane-based nanomaterial technology due to its ability to evade immune cells, thus reducing hyperreactivity. In this approach, nanoparticles serve as the core, while the membrane of red blood cells [21, 22] or platelets [23] is used to encapsulate them. This strategy offers the advantage of prolonged circulation in the bloodstream without causing cytotoxicity.

Metastasis of tumors in the brain is a common occurrence, facilitated by tumor cells crossing the BBB. Tight adhesion between endothelial cells and tumor cells is mediated by specific receptors on tumor cell surfaces and integrins and selectins on endothelial cells [24]. Therefore, the tumor cell membrane plays a crucial role in tumor metastasis.

In a study by Wang et al. in 2020, a brain metastatic tumor cell membrane-based biomimetic nanocarrier was developed, capable of crossing the BBB and facilitating imaging and PTT. The inner core of this nanocarrier consisted of indocyanine green (ICG)-loaded polymeric nanoparticles (NPs) made from poly(caprolactone) (PCL) and pluronic copolymer F68 (ICG-PCL), which were then coated with the tumor cell membrane. These ICG-loaded NPs (PCL-ICG) were encapsulated within mouse melanoma cells (B16F10), murine mammary carcinoma cancer cells (4T1), and normal cells (COS-7). Upon injection, B16-PCL-ICG and 4T1-PCL-ICG were detected in brain tumors four hours post-injection, exhibiting strong fluorescence by hour 8. Compared to naked NPs, these tumor cell membrane-camouflaged NPs demonstrated a higher ability to cross the BBB [25] (Figure 7.1).

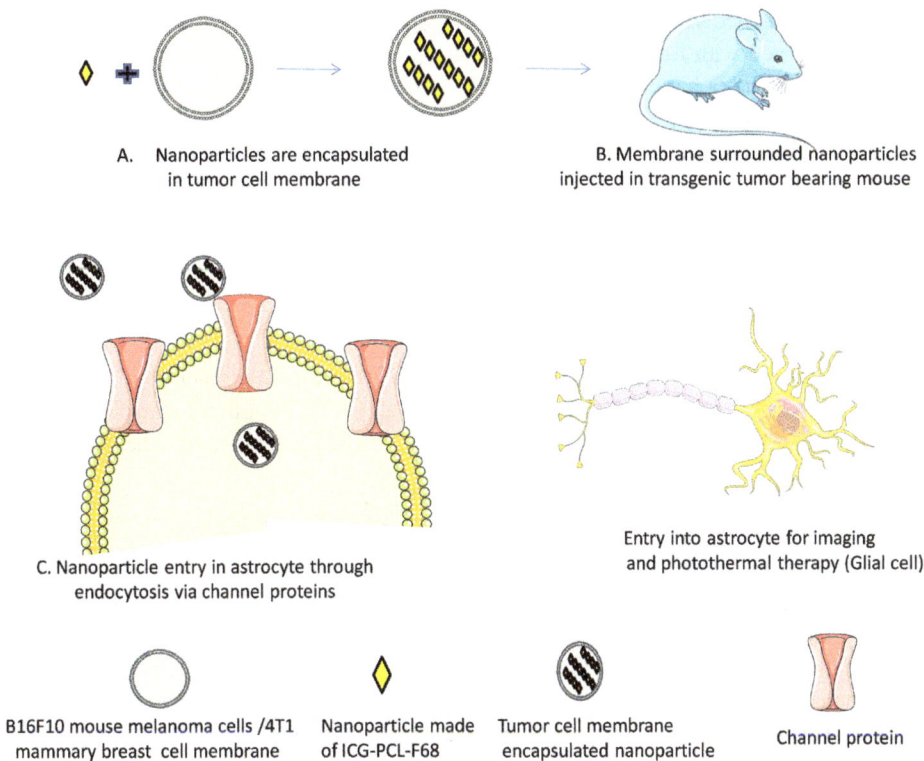

A. Nanoparticles are encapsulated in tumor cell membrane

B. Membrane surrounded nanoparticles injected in transgenic tumor bearing mouse

C. Nanoparticle entry in astrocyte through endocytosis via channel proteins

Entry into astrocyte for imaging and photothermal therapy (Glial cell)

B16F10 mouse melanoma cells /4T1 mammary breast cell membrane

Nanoparticle made of ICG-PCL-F68

Tumor cell membrane encapsulated nanoparticle

Channel protein

FIGURE 7.1 Schematic illustration of tumor cell membrane-encapsulated nanoparticles for brain tumor imaging and therapy. (A) Nanoparticles are encapsulated within the membrane of tumor cells. (B) These membrane-surrounded nanoparticles are injected into a transgenic tumor-bearing mouse. (C) The nanoparticles enter astrocytes through endocytosis via channel proteins. This process facilitates the targeted delivery of nanoparticles into glial cells for imaging and photothermal therapy.

7.5 LIQUID BIOPSY FOR BRAIN TUMORS

There are several important reasons why liquid biopsy should be considered for detecting brain tumors. First, traditional brain biopsy procedures can cause local brain swelling and bleeding around the tumor mass, posing a risk to the patient's normal neurological functions and potentially enhancing tumor aggressiveness. Additionally, focal biopsies may not fully capture the heterogeneity of the tumor tissue.

To address these challenges, integrating the analysis of various components found in the bloodstream, such as circulating tumor cells (CTCs), circulating exosomes, circulating tumor DNA (ctDNA), cell-free DNA (cfDNA), antibodies, and different metabolites, can provide a comprehensive solution for tumor analysis. Liquid biopsy offers a less invasive alternative to traditional tissue biopsies, allowing for the monitoring of tumor progression and response to treatment while minimizing risks to the patient's health.

7.5.1 CIRCULATING TUMOR CELLS

CTCs are cells shed from the primary tumor mass that enters the bloodstream, sharing similar characteristics with the primary tumor [26]. While they can contribute to metastasis, they also play a crucial role in diagnosis and prognosis. Various molecular typing methods can be employed to analyze CTCs.

One such method involves an automatic recognition algorithm based on the karyoplasmic ratio for CTC identification. CTCs isolated from the blood of glioma patients undergo DNA sequencing and immunofluorescence staining, compared with healthy controls. A low CTC count was found to be directly associated with isocitrate dehydrogenase mutations, enabling the accurate detection of oligodendroglioma with a high accuracy of 93.4% and specificity of 97.4% [27].

Diffuse gliomas, the most common primary malignant brain tumors, have been found to harbor CTCs in circulation. Despite the absence of consistently expressed markers on glioma CTCs, mutational and phenotypic stages of cancer can be determined from them. Recombinant malaria VAR2CSA protein (rVAR2) has shown promise in capturing and detecting circulating glioma cells by binding to cancer-specific oncofetal chondroitin sulfate. Targeting a panel of proteoglycans essential for tumor cell progression using rVAR2 has proven to be an effective method for targeting cancer cells [28].

In a study involving 106 glioma patients, post-operatively detected CTCs were found to be directly correlated with poor prognosis, as determined by the human telomerase reverse transcriptase assay [29]. These findings highlight the significance of CTCs in glioma diagnosis, prognosis, and therapeutic monitoring, underscoring the potential of liquid biopsy approaches in brain tumor management.

7.5.2 EXTRACELLULAR VESICLES

Extracellular vesicles (EVs), ranging in diameter from 30 to 2,000 nm, are membrane-bound vesicles carrying various biomolecules involved in cellular communication and physiological processes. Malignant cells are known to release significant quantities of EVs [30, 31], which are rich in disease-related genomic, transcriptomic, and proteomic markers [32, 33]. As such, EVs serve as a valuable source of information regarding cancer cells and their characteristics.

The growth rate of brain tumors is often slow, leading to diagnosis at a later stage, which can impact patient survival. Early detection is crucial, and, in a study by Chattrairat et al. in 2023, tumor organoid-derived EVs were analyzed for brain tumor biomarkers [34].

To detect EVs, the first step involves capturing them, and specific membrane proteins like CD9, CD63, and CD31 can be utilized for this purpose [35]. An innovative nanowire-based method was developed to capture and analyze EVs simultaneously. Zinc oxide (ZnO) nanowires were found to capture EVs based on their surface charge, which is influenced by the proteins expressed on EVs [36]. Nanowires with a strong positive charge demonstrated enhanced capturing efficiency, with highly positively charged nanowires showing superior EV capture compared to those with a lower charge [36]. ZnO/Al_2O_3 (core/shell) nanowires with a highly positive charge were produced to efficiently capture EVs.

Captured EV membrane protein profiling targeted CD63, a membrane protein overexpressed on EVs [37], and CD31, whose expression is associated with vasculogenesis and brain tumor grade and prognosis through epithelial-to-mesenchymal transitions [38]. The expression ratio of CD31/CD63 was found to differentiate between glioblastoma patients and healthy individuals. The results indicated that the nanowire system is an effective strategy for both capturing and detecting molecular markers related to brain tumors [34]. A schematic representation of the technique is provided in Figure 7.2.

Plasma extracellular vesicles (plEVs) analysis has emerged as a promising diagnostic technique for glioma. In a study involving a panel of 82 well-defined glioma patients, plEVs were separated using size exclusion-based chromatography, followed by ultrasensitive immune profiling using proximity extension assays. Syndecan-1 (SDC1), found within plEVs, serves as a discriminating factor between high-grade glioblastoma multiforme and low-grade glioma. These findings were further validated using quantitative polymerase chain reaction. The utilization of circulating plEVs as diagnostic agents presents an attractive approach for monitoring brain tumors, representing a significant step forward in brain tumor management [39].

7.5.3 Cell-Free DNA

In plasma, significant quantities of cfDNA are present, presenting an opportunity to detect genetic polymorphisms in cancer patients and monitor tumor occurrence and status non-invasively [40].

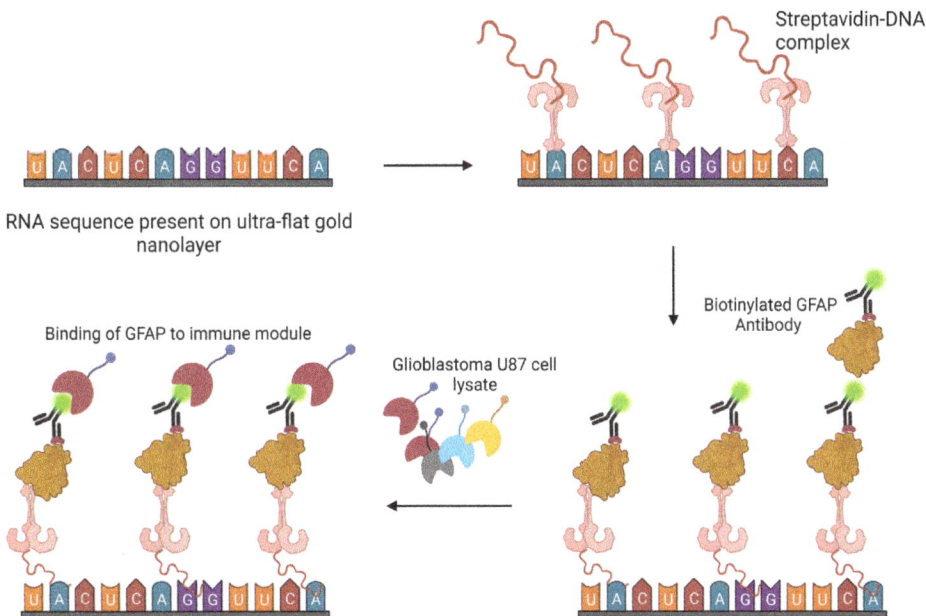

FIGURE 7.2 Schematic representation of GFAP detection using a nano-immunoassay on an ultra-flat gold nanolayer. This figure illustrates the process of detecting glial fibrillary acidic protein (GFAP) using a nano-immunoassay. An RNA sequence is present on an ultra-flat gold nanolayer (bottom). Streptavidin–DNA complexes are attached to this RNA sequence (top right). Biotinylated GFAP antibodies bind to these complexes, facilitating the capture of GFAP from glioblastoma U87 cell lysate (middle). The binding of GFAP to the immune module is shown, highlighting the detection mechanism (bottom left and center).

However, since tumor DNA comprises only a small fraction of total cfDNA, sensitive detection methods are necessary. In a study by Janku et al. in 2017, target genes were enriched using 80-mer biotinylated DNA probes spanning 61 genes. This approach revealed that personalized therapeutic interventions could be developed based on cfDNA mutation statuses [41].

To detect intracranial tumors, cfDNA from plasma can be recovered through immunoprecipitation and profiled using high-throughput sequencing (cfMeDIP-seq). A cohort of 60 diffuse glioma patients was analyzed along with published cfMeDIP-seq data. Furthermore, machine learning techniques were employed to classify gliomas among other cancerous and healthy patients [42].

7.5.4 CIRCULATING TUMOR DNA

In pediatric solid tumor patients, ctDNA has been detected in plasma [4]. Interestingly, ctDNA is found in higher concentrations in cerebrospinal fluid (CSF) compared to plasma or serum in both pediatric and adult central nervous system tumors [43]. Studies have shown that the concentration of ctDNA in CSF increases with tumor progression and is correlated with disease burden.

In a study by Miao et al. in 2021, a method was developed for detecting ctDNA mutations in diffuse intrinsic pontine gliomas (DIPGs), which are high-grade and often fatal glial tumors. DIPGs present a challenge for detection due to their deep-seated location in the brain. The technique involved a combination of cyclic enzymatic DNA amplification and surface-enhanced Raman scattering technology. A fluorescent dye Cy5-labeled probe DNA was designed to complement the mutant H3.3 DNA, forming a blunt end targetable by Exonuclease III (Exo III). The activity of Exo III led to the accumulation of large quantities of residual DNA, enhancing sensitivity even in the presence of small amounts of tumor DNA [44].

7.5.5 ANTIBODY-MEDIATED DETECTION

Glioblastoma multiforme (GBM) cells release abundant exosomes containing biomarkers that can aid in detection. A novel sensor chip has been developed, consisting of self-assembled silver NPs decorated on gold nano-islands (Ag@AuNIs), enabling the detection of exosome antigens through biotinylated antibody-mediated capture. This system detects CD63, an exosomal marker, and monocarboxylate transporter 4 (MCT4), a biomarker for GBM progression, using surface plasmon resonance (LSPR) technology. This approach offers a minimally invasive method for quantifying MCT4 in patient blood and monitoring GBM [45].

Additionally, magnetic NPs can be utilized to detect cancer cells. Iron oxide NPs, functionalized with specific antibodies targeting the epidermal growth factor receptor deletion mutant (EGFRvIII) found in human GBM cells, enhance MRI contrast in experimental GBM both in vitro and in vivo [46]. This approach shows promise for improving the visualization of GBM cells using MRI.

7.6 IMMUNOASSAY

Glial fibrillary acidic protein (GFAP) is a protein that exhibits differential expression in gliomas compared to normal brain cells. It plays a crucial role in cell differentiation, with only well-differentiated astrocytes expressing it, while aggressive tumor cells tend to lose GFAP during the de-differentiation process. To detect GFAP expression, specific immunodetection platforms have been developed, utilizing microwells capable of accommodating only one cell per microwell.

In this method, anti-GFAP antibodies are immobilized in the microwells using a DNA-streptavidin conjugate as a linker, followed by the conjugation of biotinylated monoclonal GFAP antibodies to them. The microwells are then blocked using a blocking buffer containing 3% BSA. The resulting array is tested for its ability to detect GFAP expression. Glioblastoma U87 cell lysate is incubated in the array, followed by atomic force microscopy-based imaging for visualization [47]. A schematic representation of the experimental setup is provided in Figure 7.3.

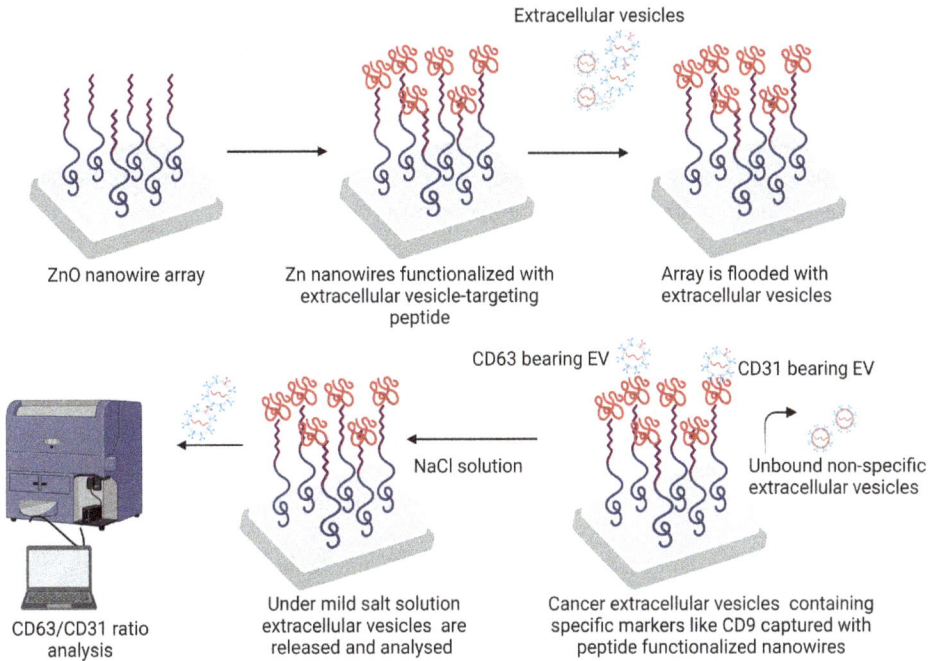

FIGURE 7.3 ZnO nanowire array functionalization and extracellular vesicle capture for brain tumor detection. This figure demonstrates the process of using a ZnO nanowire array functionalized with extracellular vesicle (EV)-targeting peptides to capture and analyze EVs for brain tumor detection. The steps include preparation of the ZnO nanowire array, functionalization of nanowires with EV-targeting peptides, flooding the array with EVs, capturing EVs bearing specific markers like CD63 and CD31 while non-specific EVs remain unbound, releasing captured EVs using a mild salt solution, and analyzing the CD63/CD31 ratio to differentiate cancerous EVs from healthy ones.

7.7 AUTOCATALYTIC NANOPARTICLE-MEDIATED IMAGING

Crossing the BBB to diagnose brain tumors poses a significant challenge for brain tumor-targeting molecules. However, gold NPs have shown promise in carrying therapeutic siRNA molecules to brain tumors for therapeutic purposes, as they can prevent the nuclease degradation of siRNA [48]. Additionally, other NPs have been explored for this purpose.

For instance, an iron oxide NP coated with a biocompatible polyethylene glycol-grafted chitosan copolymer, functionalized with chlorotoxin (CTX) and a near-infrared fluorophore, has been developed. This NP has demonstrated the ability to cross the BBB, enabling magnetic resonance and biophotonic imaging in animal models. Moreover, these nanoprobes have shown sustained retention in the tumor, facilitating the delivery of therapeutic molecules to the tumor site. This approach has opened up possibilities for flexible conjugation chemistry, allowing for the development of alternative diagnostic and therapeutic modalities [49].

A novel autocatalytic brain tumor-targeting (ABTT) delivery strategy has been proposed to improve the efficacy of NP-based therapies for brain tumors. In this approach, a small fraction of NPs initially penetrate the BBB and enter the brain tumor. Once inside the tumor environment, these NPs release BBB modulators, which transiently enhance the permeability of the BBB. This enhanced permeability allows more NPs to enter the tumor, creating a positive feedback loop system. Autocatalytic ABTT NPs were synthesized using biodegradable poly(amine-co-ester) terpolymer, ensuring biocompatibility and controlled release of BBB modulators. A high-affinity peptide derived from CTX was incorporated into the NPs, enabling specific binding to matrix metalloproteinase-2

(MMP-2), an overexpressed protein in tumors. Additionally, Lexiscan, a pharmacological agent, was utilized to transiently enhance BBB permeability and further enhance the autocatalytic activity of the NPs.

The resulting NP formulation has the potential to be modified for brain cancer imaging through PET, offering a non-invasive means of visualizing brain tumors and monitoring treatment response [50–52]. This innovative ABTT strategy represents a promising approach to improve the delivery of therapeutic agents to brain tumors while minimizing off-target effects.

7.8 CONCLUSION AND FUTURE PERSPECTIVE

Brain cancer is often diagnosed at an advanced stage, leading to poor outcomes. Thus, there is a critical need for efficient detection methods that can non-invasively detect brain tumors, ideally as part of routine check-ups. Nanoparticles have ushered in a new era of bioimaging and hold immense potential in personalized medicine. This chapter explores nanostructure-assisted fluorescence and magnetic resonance-based detection systems. One promising approach involves plasma membrane-camouflaged detecting systems, which enhance imaging and facilitate the delivery of therapeutic molecules.

Liquid biopsy offers a non-invasive method for detecting brain tumors, whether through fine needle aspirates from the affected area or CSF. Various components in clinical specimens, such as exosomes, EVs, cfDNA, ctDNA, and antibodies, can serve as biomarkers or biomarker sources. Immunoassays can be employed to detect surface antigens on exosomes, providing further diagnostic insights.

The chapter also delves into autocatalytic NP-mediated imaging techniques. Overall, nanoparticle-based brain tumor detection technologies are efficient and patient-friendly. As nanomedicine continues to evolve, it holds significant promise for improving both detection and therapy outcomes in the realm of brain cancer.

REFERENCES

[1] Torp, S.H.; Solheim, O.; Skjulsvik, A.J. The WHO 2021 Classification of Central Nervous System Tumours: A Practical Update on What Neurosurgeons Need to Know-a Minireview. *Acta Neurochir (Wien)*, 2022, *164*, 2453–2464.

[2] D'Antonio, C.; Passaro, A.; Gori, B.; Del Signore, E.; Migliorino, M.R.; Ricciardi, S.; Fulvi, A.; de Marinis, F. Bone and Brain Metastasis in Lung Cancer: Recent Advances in Therapeutic Strategies. *Ther Adv Med Oncol*, 2014, *6*, 101–114.

[3] Leone, J.P.; Leone, B.A. Breast Cancer Brain Metastases: The Last Frontier. *Exp Hematol Oncol*, 2015, *4*, 33.

[4] Liu, A.P.-Y.; Northcott, P.A.; Robinson, G.W.; Gajjar, A. Circulating Tumor DNA Profiling for Childhood Brain Tumors: Technical Challenges and Evidence for Utility. *Lab Invest*, 2022, *102*, 134–142.

[5] Foster, D.; Larsen, J. Polymeric Metal Contrast Agents for T1-Weighted Magnetic Resonance Imaging of the Brain. *ACS Biomater. Sci. Eng.*, 2023, *9*, 1224–1242.

[6] Puskar, A.; Saadah, B.; Rauf, A.; Kasperek, S.R.; Umair, M. A Primer on Contrast Agents for Magnetic Resonance Imaging of Post-Procedural and Follow-Up Imaging of Islet Cell Transplant. *Nano Select*, 2023, *4*, 181–191.

[7] Farooqi, S.; Mumtaz, A.; Arif, A.; Butt, M.; Kanor, U.; Memoh, S.; Qamar, M.A.; Yosufi, A. The Clinical Manifestations and Efficacy of Different Treatments Used for Nephrogenic Systemic Fibrosis: A Systematic Review. *Int J Nephrol Renovasc Dis*, 2023, *16*, 17–30.

[8] Saeedi, J.A.; AlYafeai, R.H.; AlAbdulSalam, A.M.; Al-Dihan, A.Y.; AlDwaihi, A.A.; Al Harbi, A.A.; Aljadhai, Y.I.; Al-Jedai, A.H.; AlKhawajah, N.M.; Al-Luqmani, M.M.; AlMalki, A.O.; Al-Mudaiheem, H.Y.; AlNajashi, H.A.; AlShareef, R.A.; AlShehri, A.A.; AlThekair, F.Y.; Ben Slimane, N.S.; Cupler, E.J.; Kalakatawi, M.H.; Kedah, H.M.; Al Malik, Y.M.; Althubaiti, I.A.; Bunyan, R.F.; Shosha, E.; Al Jumah, M.A. Saudi Consensus Recommendations on the Management of Multiple Sclerosis: Diagnosis and Radiology/Imaging. *Clinical and Translational Neuroscience*, 2023, *7*, 5.

[9] Chan, M.-H.; Chen, W.; Li, C.-H.; Fang, C.-Y.; Chang, Y.-C.; Wei, D.-H.; Liu, R.-S.; Hsiao, M. An Advanced In Situ Magnetic Resonance Imaging and Ultrasonic Theranostics Nanocomposite Platform: Crossing the Blood-Brain Barrier and Improving the Suppression of Glioblastoma Using Iron-Platinum Nanoparticles in Nanobubbles. *ACS Appl Mater Interfaces*, 2021, *13*, 26759–26769.

[10] Duan, Y.; Hu, D.; Guo, B.; Shi, Q.; Wu, M.; Xu, S.; Kenry; Liu, X.; Jiang, J.; Sheng, Z.; Zheng, H.; Liu, B. Nanostructural Control Enables Optimized Photoacoustic–Fluorescence–Magnetic Resonance Multimodal Imaging and Photothermal Therapy of Brain Tumor. *Adv Funct Mater*, 2020, *30*, 1907077.

[11] Pabian-Jewuła, S.; Bragiel-Pieczonka, A.; Rylski, M. Ying Yang 1 Engagement in Brain Pathology. *J Neurochem*, 2022, *161*, 236–253.

[12] Singh, P.; Yang, M.; Dai, H.; Yu, D.; Huang, Q.; Tan, W.; Kernstine, K.H.; Lin, D.; Shen, B. Overexpression and Hypomethylation of Flap Endonuclease 1 Gene in Breast and Other Cancers. *Mol Cancer Res*, 2008, *6*, 1710–1717.

[13] Li, B.; Zhang, P.; Zhou, B.; Xie, S.; Xia, A.; Suo, T.; Feng, S.; Zhang, X. Fluorometric Detection of Cancer Marker FEN1 Based on Double-Flapped Dumbbell DNA Nanoprobe Functionalized with Silver Nanoclusters. *Anal Chim Acta*, 2021, *1148*, 238194.

[14] Kircher, M.F.; de la Zerda, A.; Jokerst, J.V.; Zavaleta, C.L.; Kempen, P.J.; Mittra, E.; Pitter, K.; Huang, R.; Campos, C.; Habte, F.; Sinclair, R.; Brennan, C.W.; Mellinghoff, I.K.; Holland, E.C.; Gambhir, S.S. A Brain Tumor Molecular Imaging Strategy Using a New Triple-Modality MRI-Photoacoustic-Raman Nanoparticle. *Nat Med*, 2012, *18*, 829–834.

[15] Zheng, M.; Ruan, S.; Liu, S.; Sun, T.; Qu, D.; Zhao, H.; Xie, Z.; Gao, H.; Jing, X.; Sun, Z. Self-Targeting Fluorescent Carbon Dots for Diagnosis of Brain Cancer Cells. *ACS Nano*, 2015, *9*, 11455–11461.

[16] Duan, Y.; Zhang, C.; Wang, Y.; Chen, G. Research Progress of Whole-Cell-SELEX Selection and the Application of Cell-Targeting Aptamer. *Mol Biol Rep*, 2022, *49*, 7979–7993.

[17] Mohammadi, R.; Naderi-Manesh, H.; Farzin, L.; Vaezi, Z.; Ayarri, N.; Samandari, L.; Shamsipur, M. Fluorescence Sensing and Imaging with Carbon-Based Quantum Dots for Early Diagnosis of Cancer: A Review. *J Pharm Biomed Anal*, 2022, *212*, 114628.

[18] Chakraborty, P.; Das, S.S.; Dey, A.; Chakraborty, A.; Bhattacharyya, C.; Kandimalla, R.; Mukherjee, B.; Gopalakrishnan, A.V.; Singh, S.K.; Kant, S.; Nand, P.; Ojha, S.; Kumar, P.; Jha, N.K.; Jha, S.K.; Dewanjee, S. Quantum Dots: The Cutting-Edge Nanotheranostics in Brain Cancer Management. *J Control Release*, 2022, *350*, 698–715.

[19] Meyer, E.L.; Mbese, J.Z.; Agoro, M.A. The Frontiers of Nanomaterials (SnS, PbS and CuS) for Dye-Sensitized Solar Cell Applications: An Exciting New Infrared Material. *Molecules*, 2019, *24*, 4223.

[20] Barik, P.; Pradhan, M. All-Optical Detection of Biocompatible Quantum Dots. In: *Application of Quantum Dots in Biology and Medicine: Recent Advances*; Barik, P.; Mondal, S., Eds.; Springer Nature: Singapore, 2022; pp. 35–65.

[21] Li, J.; Wei, Y.; Zhang, C.; Bi, R.; Qiu, Y.; Li, Y.; Hu, B. Cell-Membrane-Coated Nanoparticles for Targeted Drug Delivery to the Brain for the Treatment of Neurological Diseases. *Pharmaceutics*, 2023, *15*, 621.

[22] Zafar, H.; Yousefiasl, S.; Raza, F. T-Cell Membrane-Functionalized Nanosystems for Viral Infectious Diseases. *Mater Chem Horiz*, 2023, *2*, 41–48.

[23] Liu, H.; Su, Y.-Y.; Jiang, X.-C.; Gao, J.-Q. Cell Membrane-Coated Nanoparticles: A Novel Multifunctional Biomimetic Drug Delivery System. *Drug Deliv Transl Res*, 2023, *13*, 716–737.

[24] Ludwig, B.S.; Kessler, H.; Kossatz, S.; Reuning, U. RGD-Binding Integrins Revisited: How Recently Discovered Functions and Novel Synthetic Ligands (Re-)Shape an Ever-Evolving Field. *Cancers*, 2021, *13*, 1711.

[25] Wang, C.; Wu, B.; Wu, Y.; Song, X.; Zhang, S.; Liu, Z. Camouflaging Nanoparticles with Brain Metastatic Tumor Cell Membranes: A New Strategy to Traverse Blood–Brain Barrier for Imaging and Therapy of Brain Tumors. *Adv Funct Mater*, 2020, *30*, 1909369.

[26] Klotz, R.; Yu, M. Insights into Brain Metastasis: Recent Advances in Circulating Tumor Cell Research. *Cancer Rep (Hoboken)*, 2022, *5*, e1239.

[27] Zhu, X.; Wen, S.; Deng, S.; Wu, G.; Tian, R.; Hu, P.; Ye, L.; Sun, Q.; Xu, Y.; Deng, G.; Zhang, D.; Yang, S.; Qi, Y.; Chen, Q. A Novel Karyoplasmic Ratio-Based Automatic Recognition Method for Identifying Glioma Circulating Tumor Cells. *Front Oncol*, 2022, *12*, 893769.

[28] Bang-Christensen, S.R.; Pedersen, R.S.; Pereira, M.A.; Clausen, T.M.; Løppke, C.; Sand, N.T.; Ahrens, T.D.; Jørgensen, A.M.; Lim, Y.C.; Goksøyr, L.; Choudhary, S.; Gustavsson, T.; Dagil, R.; Daugaard, M.; Sander, A.F.; Torp, M.H.; Søgaard, M.; Theander, T.G.; Østrup, O.; Lassen, U.; Hamerlik, P.; Salanti, A.; Agerbæk, M.Ø. Capture and Detection of Circulating Glioma Cells Using the Recombinant VAR2CSA Malaria Protein. *Cells*, 2019, *8*, 998.

[29] Zhang, W.; Qin, T.; Yang, Z.; Yin, L.; Zhao, C.; Feng, L.; Lin, S.; Liu, B.; Cheng, S.; Zhang, K. Telomerase-Positive Circulating Tumor Cells Are Associated with Poor Prognosis via a Neutrophil-Mediated Inflammatory Immune Environment in Glioma. *BMC Med*, 2021, *19*, 277.

[30] Shojaei, S.; Hashemi, S.M.; Ghanbarian, H.; Salehi, M.; Mohammadi-Yeganeh, S. Effect of Mesenchymal Stem Cells-Derived Exosomes on Tumor Microenvironment: Tumor Progression versus Tumor Suppression. *J Cell Physiol*, 2019, *234*, 3394–3409.

[31] Sun, D.; Ma, Y.; Wu, M.; Chen, Z.; Zhang, L.; Lu, J. Recent Progress in Aptamer-Based Microfluidics for the Detection of Circulating Tumor Cells and Extracellular Vesicles. *Journal of Pharmaceutical Analysis*, 2023, *13*(4), 340–354.

[32] Medhin, L.B.; Beasley, A.B.; Warburton, L.; Amanuel, B.; Gray, E.S. Extracellular Vesicles as a Liquid Biopsy for Melanoma: Are We There Yet? *Semin Cancer Biol*, 2023, *89*, 92–98.

[33] Wong, L.-W.; Mak, S.-H.; Goh, B.-H.; Lee, W.-L. The Convergence of FTIR and EVs: Emergence Strategy for Non-Invasive Cancer Markers Discovery. *Diagnostics*, 2023, *13*, 22.

[34] Chattrairat, K.; Yasui, T.; Suzuki, S.; Natsume, A.; Nagashima, K.; Iida, M.; Zhang, M.; Shimada, T.; Kato, A.; Aoki, K.; Ohka, F.; Yamazaki, S.; Yanagida, T.; Baba, Y. All-in-One Nanowire Assay System for Capture and Analysis of Extracellular Vesicles from an Ex Vivo Brain Tumor Model. *ACS Nano*, 2023, *17*, 2235–2244.

[35] Dixson, A.C.; Dawson, T.R.; Di Vizio, D.; Weaver, A.M. Context-Specific Regulation of Extracellular Vesicle Biogenesis and Cargo Selection. *Nat Rev Mol Cell Biol*, 2023, 1–23.

[36] Yasui, T.; Paisrisarn, P.; Yanagida, T.; Konakade, Y.; Nakamura, Y.; Nagashima, K.; Musa, M.; Thiodorus, I.A.; Takahashi, H.; Naganawa, T.; Shimada, T.; Kaji, N.; Ochiya, T.; Kawai, T.; Baba, Y. Molecular Profiling of Extracellular Vesicles via Charge-Based Capture Using Oxide Nanowire Microfluidics. *Biosens Bioelectron*, 2021, *194*, 113589.

[37] Wang, R.; Wang, X.; Zhang, Y.; Zhao, H.; Cui, J.; Li, J.; Di, L. Emerging Prospects of Extracellular Vesicles for Brain Disease Theranostics. *J Control Release*, 2022, *341*, 844–868.

[38] Fahmy, S.A.; Dawoud, A.; Zeinelabdeen, Y.A.; Kiriacos, C.J.; Daniel, K.A.; Eltahtawy, O.; Abdelhalim, M.M.; Braoudaki, M.; Youness, R.A. Molecular Engines, Therapeutic Targets, and Challenges in Pediatric Brain Tumors: A Special Emphasis on Hydrogen Sulfide and RNA-Based Nano-Delivery. *Cancers (Basel)*, 2022, *14*, 5244.

[39] Indira Chandran, V.; Welinder, C.; Månsson, A.-S.; Offer, S.; Freyhult, E.; Pernemalm, M.; Lund, S.M.; Pedersen, S.; Lehtiö, J.; Marko-Varga, G.; Johansson, M.C.; Englund, E.; Sundgren, P.C.; Belting, M. Ultrasensitive Immunoprofiling of Plasma Extracellular Vesicles Identifies Syndecan-1 as a Potential Tool for Minimally Invasive Diagnosis of Glioma. *Clin Cancer Res*, 2019, *25*, 3115–3127.

[40] Chen, Y.; Gong, Y.; Dou, L.; Zhou, X.; Zhang, Y. Bioinformatics Analysis Methods for Cell-Free DNA. *Comput Biol Med*, 2022, *143*, 105283.

[41] Janku, F.; Zhang, S.; Waters, J.; Liu, L.; Huang, H.J.; Subbiah, V.; Hong, D.S.; Karp, D.D.; Fu, S.; Cai, X.; Ramzanali, N.M.; Madwani, K.; Cabrilo, G.; Andrews, D.L.; Zhao, Y.; Javle, M.; Kopetz, E.S.; Luthra, R.; Kim, H.J.; Gnerre, S.; Satya, R.V.; Chuang, H.-Y.; Kruglyak, K.M.; Toung, J.; Zhao, C.; Shen, R.; Heymach, J.V.; Meric-Bernstam, F.; Mills, G.B.; Fan, J.-B.; Salathia, N.S. Development and Validation of an Ultradeep Next-Generation Sequencing Assay for Testing of Plasma Cell-Free DNA from Patients with Advanced Cancer. *Clin Cancer Res*, 2017, *23*, 5648–5656.

[42] Nassiri, F.; Chakravarthy, A.; Feng, S.; Shen, S.Y.; Nejad, R.; Zuccato, J.A.; Voisin, M.R.; Patil, V.; Horbinski, C.; Aldape, K.; Zadeh, G.; De Carvalho, D.D. Detection and Discrimination of Intracranial Tumors Using Plasma Cell-Free DNA Methylomes. *Nat Med*, 2020, *26*, 1044–1047.

[43] Wang, Y.; Springer, S.; Zhang, M.; McMahon, K.W.; Kinde, I.; Dobbyn, L.; Ptak, J.; Brem, H.; Chaichana, K.; Gallia, G.L.; Gokaslan, Z.L.; Groves, M.L.; Jallo, G.I.; Lim, M.; Olivi, A.; Quinones-Hinojosa, A.; Rigamonti, D.; Riggins, G.J.; Sciubba, D.M.; Weingart, J.D.; Wolinsky, J.-P.; Ye, X.; Oba-Shinjo, S.M.; Marie, S.K.N.; Holdhoff, M.; Agrawal, N.; Diaz, L.A.; Papadopoulos, N.; Kinzler, K.W.; Vogelstein, B.; Bettegowda, C. Detection of Tumor-Derived DNA in Cerebrospinal Fluid of Patients with Primary Tumors of the Brain and Spinal Cord. *Proc Natl Acad Sci U S A*, 2015, *112*, 9704–9709.

[44] Miao, X.; Fang, Q.; Xiao, X.; Liu, S.; Wu, R.; Yan, J.; Nie, B.; Liu, J. Integrating Cycled Enzymatic DNA Amplification and Surface-Enhanced Raman Scattering for Sensitive Detection of Circulating Tumor DNA. *Front Mol Biosci*, 2021, *8*, 676065.

[45] Liu, L.; Thakur, A.; Kar Li, W.; Qiu, G.; Yang, T.; He, B.; Lee, Y.; Lawrence Wu, C.-M. Site Specific Biotinylated Antibody Functionalized Ag@AuNIs LSPR Biosensor for the Ultrasensitive Detection of Exosomal MCT4, a Glioblastoma Progression Biomarker. *Chem Eng J*, 2022, *446*, 137383.

[46] Hadjipanayis, C.G.; Machaidze, R.; Kaluzova, M.; Wang, L.; Schuette, A.J.; Chen, H.; Wu, X.; Mao, H. EGFRvIII Antibody-Conjugated Iron Oxide Nanoparticles for Magnetic Resonance Imaging-Guided Convection-Enhanced Delivery and Targeted Therapy of Glioblastoma. *Cancer Res*, 2010, *70*, 6303–6312.

[47] Ganau, M.; Bosco, A.; Palma, A.; Corvaglia, S.; Parisse, P.; Fruk, L.; Beltrami, A.P.; Cesselli, D.; Casalis, L.; Scoles, G. A DNA-Based Nano-Immunoassay for the Label-Free Detection of Glial Fibrillary Acidic Protein in Multicell Lysates. *Nanomedicine*, 2015, *11*, 293–300.

[48] Jensen, S.A.; Day, E.S.; Ko, C.H.; Hurley, L.A.; Luciano, J.P.; Kouri, F.M.; Merkel, T.J.; Luthi, A.J.; Patel, P.C.; Cutler, J.I.; Daniel, W.L.; Scott, A.W.; Rotz, M.W.; Meade, T.J.; Giljohann, D.A.; Mirkin, C.A.; Stegh, A.H. Spherical Nucleic Acid Nanoparticle Conjugates as an RNAi-Based Therapy for Glioblastoma. *Sci Transl Med*, 2013, *5*, 209ra152.

[49] Veiseh, O.; Sun, C.; Fang, C.; Bhattarai, N.; Gunn, J.; Kievit, F.; Du, K.; Pullar, B.; Lee, D.; Ellenbogen, R.G.; Olson, J.; Zhang, M. Specific Targeting of Brain Tumors with an Optical/Magnetic Resonance Imaging Nanoprobe across the Blood-Brain Barrier. *Cancer Res*, 2009, *69*, 6200–6207.

[50] Li, M.; Yan, T.; Cai, Y.; Wei, Y.; Xie, Q. Expression of Matrix Metalloproteinases and Their Association with Clinical Characteristics of Solid Tumors. *Gene*, 2023, *850*, 146927.

[51] Fukuta, T.; Oku, N.; Kogure, K. Application and Utility of Liposomal Neuroprotective Agents and Biomimetic Nanoparticles for the Treatment of Ischemic Stroke. *Pharmaceutics*, 2022, *14*, 361.

[52] Han, L.; Kong, D.K.; Zheng, M.-Q.; Murikinati, S.; Ma, C.; Yuan, P.; Li, L.; Tian, D.; Cai, Q.; Ye, C.; Holden, D.; Park, J.-H.; Gao, X.; Thomas, J.-L.; Grutzendler, J.; Carson, R.E.; Huang, Y.; Piepmeier, J.M.; Zhou, J. Increased Nanoparticle Delivery to Brain Tumors by Autocatalytic Priming for Improved Treatment and Imaging. *ACS Nano*, 2016, *10*, 4209–4218.

8 Innovative Approaches in Brain Cancer Therapy

Nanomedicine and Targeted Delivery Systems

Yawen Ma[1]*, Rekha Khandia[2], and Pankaj Gurjar[3]

[1]Department of Neurosurgery, Shengjing Hospital of China Medical University, Shenyang, Liaoning, China

[2]Department of Biochemistry and Genetics, Barkatullah University, Bhopal MP, India

[3]Department of Science and Engineering, Novel Global Community Educational Foundation, Hebersham, NSW, Australia

*Corresponding author

8.1 INTRODUCTION

Malignant brain tumors are rare in adults, accounting for only 1–2% of all cancers [1]. Despite their rarity, they are highly deadly, necessitating new treatment strategies. These tumors typically arise in glial cells and are graded by their aggressiveness, with grade IV being the most severe. Glioblastoma multiforme (GBM) is the most common and aggressive type, with a median survival of less than two years [2]. The challenge in treating brain tumors lies in their location within the brain, where surgical removal is often difficult. Additionally, the blood–brain barrier (BBB) makes delivering effective chemotherapy agents challenging. The BBB tightly controls the passage of molecules from the bloodstream to the central nervous system, complicating the delivery of drugs meant to target tumors. Moreover, the brain's immunosuppressive nature, genetic instability promoting tumor growth, and the complex variability within and between tumors make traditional treatment approaches inefficient. This often leads to the use of high doses of chemotherapy or radiotherapy, which can cause severe side effects and reduce quality of life. The main hurdles to effective treatment are the BBB's limited penetrability and drug resistance caused by an efficient efflux system. These challenges underscore the need for alternative therapeutic approaches. In this chapter, we explore innovative strategies for treating brain cancer, including the development of nanomedicines that have transitioned from laboratory research to clinical trials. We discuss the latest advancements and the ongoing challenges in developing effective therapies for brain tumors.

8.2 NANOMEDICINES

Recent advancements in nanotechnology have brought a revolution in various scientific fields, including medicine. In the realm of medical science, innovative strategies are emerging to combat brain cancer, particularly GBM. These strategies offer alternatives or complementary approaches to traditional tumor treatments [3–9]. One promising avenue is drug delivery via nanoparticles (NPs), which can encapsulate or modify biological molecules for targeted therapy. Several research groups are actively developing cytotoxic NPs, with some already in the approval process by the Food and Drug Administration (FDA) to minimize side effects. NPs can be synthesized using different

DOI: 10.1201/9781003519706-8

chemical or environmentally friendly methods [10–16]. For instance, paclitaxel is a well-known anticancer drug. In a study by Wiwatchaitawee et al. in 2022, NP-mediated delivery of paclitaxel was investigated. Paclitaxel was encapsulated in poly(lactic-co-glycolic acid) (PLGA)–polyethylene glycol (PEG) NPs, which were surface-modified with positively charged molecules. This modification facilitated significant accumulation of the NPs within tumor cells. Following treatment, an increase in survival was observed in mice with GBM xenografts. These findings underscore the importance of surface charge in NP-mediated delivery and retention of therapeutic molecules within the targeted organ (Figure 8.1) [17].

Poly-lactic acid NPs (PLA-NP) are naturally hydrophobic, making them prone to uptake by macrophages and the reticuloendothelial system. To enhance their circulation time, they can be coated with hydrophilic stabilizers. These surface-modified PLA NPs can then be functionalized with temozolomide (TMZ), an anticancer agent. Studies have shown that such NPs can effectively cross the BBB in C6 glioma cells, leading to improved drug half-life and mean residence time [18]. However, because NPs have limited sensitivity toward cancer cells, they may inadvertently accumulate in other tissues, causing injury to normal cells [19]. To address this issue, nanomedicines can be functionalized with ligands that selectively bind to receptors expressed or overexpressed on tumor tissues [20–22]. Examples of such receptors include folate, transferrin, neurokinin-1, and v3 integrin receptors [23–25]. Albumin NPs have shown promise in efficiently delivering therapeutic molecules to solid tumors. For example, one-pot synthesis of doxorubicin-loaded naive albumin NPs and mannose or cationic-modified albumin NPs demonstrated stability and extended release of doxorubicin in U87MG glioblastoma cells and spheroids. These formulations can serve as effective anti-tumor agents against glioma, offering sustained release stability and targetability [26]. Similarly, doxorubicin and paclitaxel-loaded transferrin-conjugated magnetic silica PLGA NPs (MNP–MSN–PLGA–Tf NPs) have been formulated. These NPs exhibit enhanced cellular uptake when a magnetic field is applied, leading to superior cell toxicity in U-87 cells. In an intracranial U-87 MG-luc2 xenograft model in BALB/c nude mice, these nanoformulations demonstrated anti-glioma activity, representing promising drugs for treating solid brain tumors [27]. Brain metastasis is reported in around 20% of systemic cancers [28] and contributes significantly to cancer-related deaths with poor prognoses. Among all cancer types, breast cancer is the one that commonly metastasizes to the brain (breast cancer brain metastasis [BCBM]) and is associated with the poorest survival with a median time

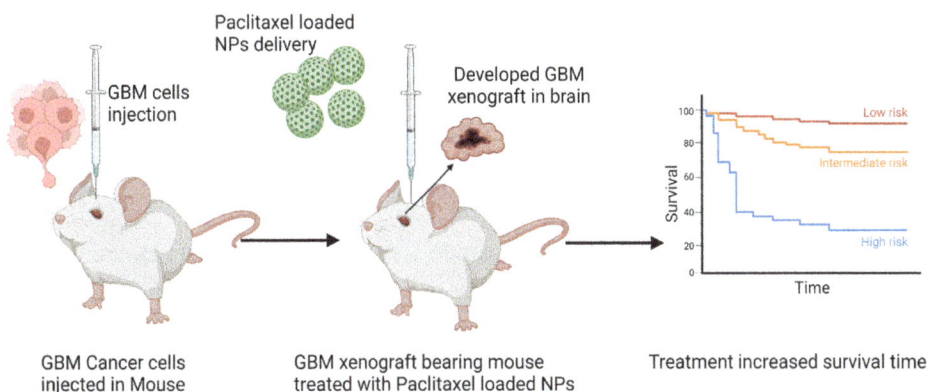

FIGURE 8.1 Treatment of glioblastoma multiforme (GBM) xenografts in mice using paclitaxel-loaded nanoparticles. This figure demonstrates the process of treating GBM xenografts in mice using paclitaxel-loaded nanoparticles. The sequence includes the injection of GBM cancer cells into a mouse to develop a brain xenograft, the administration of paclitaxel-loaded nanoparticles to the GBM-bearing mouse, and the assessment of survival rates. The survival rates are categorized by risk levels (high, intermediate, low) and the treatment shows an increased survival time in treated mice.

of four to six months. The poor prognosis is the outcome of insufficient drug delivery to the metastatic lesions in the brain. Improvements in therapeutic molecules so that these may pass through the blood–brain tumor barrier (BTB) can improve the drug's efficacy. Prostate-specific membrane antigen (PSMA) overexpress in the neovasculature of many cancers and is a transmembrane protein helping in receptor-mediated endocytosis. It may help in internalizing the PSMA ligands across the cell.

Furthermore, PSMA is known to be overexpressed on the surface of blood–cerebrospinal fluid barrier (BCB) endothelial cells associated with BCBM but not on typical BBB endothelial cells. This unique expression pattern has been exploited for targeted therapy using doxorubicin (DOX) and lapatinib drug-functionalized NPs, which have shown efficacy in treating these metastases [29]. By targeting PSMA, the side effects associated with doxorubicin therapy can be minimized through more selective drug delivery.

Transferrin receptor-mediated endocytosis presents another promising avenue for targeted drug delivery across the BBB [30]. Carbon dots (C-dots) exhibit favorable biocompatibility, solubility, stability, and surface properties, along with excellent photoluminescence. In a study by Hettiarachchi et al., transferrin and two anticancer drugs, epirubicin and TMZ, were conjugated with CDs sized between 1.5 and 1.7 nm, resulting in particles with a size of 3.5 nm. These transferrin conjugates demonstrated potent anti-tumor effects, with low tumor cell viability even at lower concentrations, achieving 86% cytotoxicity in SJGBM2 cell lines [31].

Moreover, non-toxic carbon nitride dots (CNDs) have been engineered to selectively target pediatric glioblastoma cells. These CNDs, conjugated with transferrin and gemcitabine, a chemotherapeutic drug, exhibited selective toxicity against SJGBM2 glioma cells while leaving healthy cells unharmed, with 100% viability even at a low concentration of 10 nM. These findings were further validated in a zebrafish model, highlighting the remarkable nanocarrier capabilities of CNDs [32].

Hyaluronic acid (HA) is a negatively charged molecule known to bind to specific receptors that are often upregulated in various cancers, including cluster determinant 44 (CD44) receptor, the HA-mediated motility receptor, and lymphatic endothelial-1. This property makes HA a promising targeting molecule for cancer therapy [33]. HA-modified gold nanocages (AuNCs-HA) have been developed that are ideal for photoacoustic imaging and combined photothermal and radiotherapy therapy. In vivo, imaging results demonstrated the developed system provided photoacoustic imaging of cancer's contour, size, and location and enhanced tumor cell killing via radio and Photothermal therapy [34].

To combat aggressive brain tumors, novel theranostic platforms are being developed. One such platform is a nanoplatform designed to respond to changes in pH, glutathione (GSH) levels, and hyaluronidase activity. This platform is targeted against HER2 and CD44 antigens and can efficiently deliver both photodynamic therapy (PDT) and photothermal therapy. With a circulation half-life of 1.9 hours, this versatile system, when coupled with an image-guiding system, has shown promise in eliminating tumors without apparent side effects [35]. It represents hope for scientists aiming to develop theranostics with high tumor-eliminating activity and minimal side effects for cancer patients.

Additionally, nitrogen-doped graphene oxide dots (NGODs), coupled with ascorbic acid, have been developed to produce hydrogen peroxide (H_2O_2) under white-light irradiation. This PDT approach effectively kills tumor cells by inducing apoptosis and necrosis upon internalization of NGODs into the cells [36–38].

8.3 PLANT-BASED PRODUCTS FOR BRAIN TUMOR TREATMENT

Many plant-derived products have shown promise in cancer treatment [39]. The Cannabis plant, in particular, produces numerous phytochemicals, with approximately 50 being produced in significant quantities. Among these, delta-9-tetrahydrocannabinol (THC) and cannabidiol (CBD) are the most extensively studied [40]. Both THC and CBD have been found to inhibit cancer cell migration, angiogenesis, and induce apoptosis [41].

In the body, the balance between cannabinoid receptors and their ligands plays a crucial role, which is often altered during the transition from normal to malignant cells in various cancers, including brain cancer [42]. Glioblastoma, for example, expresses both cannabinoid receptors CB1 and CB2, whereas normal cells typically only express CB1. However, during cancer development, CB2 is often overexpressed [43], making brain cancer cells susceptible to CB2 antagonists. One such antagonist is arachidonoylethanolamide, which shows promise as a therapeutic molecule against cancer [44].

Despite these advancements, the BBB poses a significant challenge by hindering the entry of therapeutic molecules into the tumor, rendering treatments less effective. However, many plant-based chemicals have been identified as potential modulators of the BBB in the tumor microenvironment (TME). These compounds offer potential solutions to improve drug delivery to brain tumors and enhance treatment efficacy.

Shikonin, a liposoluble derivative of anthraquinone derived from the root of *Lithospermum erythrorhizon*, has shown promising effects in the therapy of brain tumors [45]. It exerts its effects by increasing reactive oxygen species and Bax levels, while decreasing GSH, Bcl-2, and catalase. Additionally, Shikonin (SHK) disrupts mitochondrial transmembrane potential, upregulates p53 and SOD-1, and cleaves poly-ADP-ribose-polymerase (PARP) in glioma cell lines, demonstrating its multitarget effects on brain tumors.

When combined with TMZ, a commonly used chemotherapeutic drug, SHK has been shown to reduce proliferation and migration capacities and induce apoptosis and necroptosis in human glioma cells by inhibiting epidermal growth factor receptor signaling [46]. This combination therapy also targets glial-to-mesenchymal transition, a process responsible for tumor invasion and malignancy, by reducing the expression of β3 integrin, MMP-2, MMP-9, Slug, and vimentin [47].

Furthermore, incorporating SHK with other chemotherapeutic drugs may further enhance the therapeutic efficacy of chemotherapy regimens for brain tumors, offering a potential avenue for improving patient outcomes.

Encapsulating SHK within NPs composed of PEG–PLGA and coated with lactoferrin has been shown to enhance the ability to cross the BBB and target brain tumor cells, making it a promising option for targeted therapy of glioblastoma cells. This innovative approach holds potential for improving the efficacy of treatment for glioblastoma, a challenging cancer to treat due to its location and resistance to conventional therapies (Figure 8.2) [48].

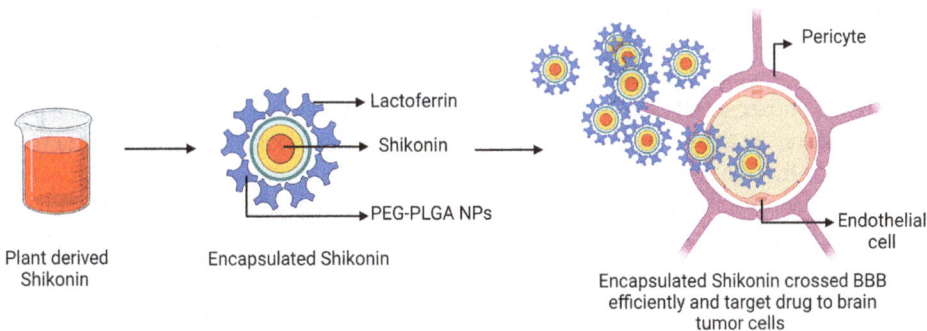

FIGURE 8.2 Encapsulation and targeted delivery of Shikonin to brain tumor cells using lactoferrin-functionalized PEG–PLGA nanoparticles. This figure demonstrates the process of encapsulating plant-derived Shikonin in PEG–PLGA nanoparticles functionalized with lactoferrin for targeted drug delivery to brain tumor cells. The steps include the extraction of Shikonin from plants, encapsulation of Shikonin in PEG–PLGA nanoparticles, functionalization of these nanoparticles with lactoferrin, and the subsequent crossing of the blood–brain barrier (BBB) to target brain tumor cells effectively.

Osthole, an active component found in cnidium fruit, has demonstrated anticancer properties by inhibiting proliferation, inducing apoptosis, and inhibiting migration and invasion in C6 glioma cells [49]. It achieves these effects by inhibiting the phosphorylation of focal adhesion kinase and reducing the expression of matrix metalloproteinase (MMP)-13, thereby suppressing glioblastoma multiforme cell motility [50]. The ethanol extract of *Securidaca longipedunculata* Fresen plant, traditionally used in Nigerian medicine, has also shown efficacy in inducing apoptosis in the U87 malignant brain tumor cell line through the cleavage of PARP [51]. Resveratrol (RES), another plant-derived compound, has been evaluated for its effects on GBM cell lines, with a primary focus on autophagy, apoptosis, and necrosis. RES was found to upregulate autophagic genes Atg5, beclin-1, and LC3-II, while altering the expression of cyclins A, E, and B, pRb, and cyclin D1 [52]. These findings suggest that RES could potentially enhance the therapeutic efficacy of other drugs when combined with them. A furanocoumarin glycoside isolated from the twigs of *Dorstenia turbinata* demonstrated inhibition of MMP secretion in brain tumors, indicating its potential as a target for therapy development. This compound showed anti-MMP activity comparable to chlorogenic acid (CHL) and epigallocatechin-3-gallate [53].

Additionally, microsomal glucose-6-phosphate translocase was found to be overexpressed in U-87 glioma cells. CHL, found in tea and known for its medicinal and antioxidant properties, has been shown to potently inhibit MMP9 activity and may serve as a promising chemotherapeutic agent [54–57]. These plant-derived compounds offer potential avenues for developing novel therapies or enhancing the efficacy of existing treatments for brain tumors.

8.4   RNA INTERFERENCE

Interfering RNA (RNAi) holds promise for knocking down cancer oncogenes; however, its therapeutic use is limited by the high degradability of RNA molecules. To overcome this limitation, plasmid DNA can be engineered to express short hairpin RNA (shRNA), which mimics endogenous microRNAs. In a study by Pardridge in 2005 [58], an RNAi-based gene therapy module was developed using plasmid DNA encapsulated within receptor-specific pegylated immunoliposomes, capable of crossing the endothelial cell barrier and tumor cell and nuclear membranes. Insulin receptors, present on the BBB of both normal and cancer cells [59] and abundant on the surface of all brain cancer cells [60], can shuttle their ligand insulin to the nucleus [61]. A peptidomimetic monoclonal antibody (mAb) designed to bind to the insulin receptor, along with the therapeutic DNA, can be utilized to transport the molecule to the nucleus. These therapeutic molecules can be entrapped within liposomes, PEGylated, and decorated with receptors to successfully deliver therapeutic molecules. This approach has shown promising results, achieving a 90% knockdown of brain tumor-specific genes with a single intravenous injection in mouse models. RNAi technology can also be coupled with the delivery of tumor suppressor genes to synergize therapeutic potential [58].

Brain tumor-initiating cells are the primary drivers of therapy resistance, recurrence, and progression in brain tumors. In a study by Yu et al. in 2017, siRNA-mediated knockdown of four transcription factors (SOX2, OLIG2, SALL2, and POU3F2) was achieved simultaneously using lipopolymeric NPs 7C1. This approach lays the groundwork for multiplexing RNAi strategies targeting various tumors, including brain tumors, to address the challenges posed by tumor heterogeneity [62]. Malignant GBM is one of the most aggressive tumors, with a mean survival of only 10–15 months even with surgery and radiotherapy. The TME in GBM is highly heterogeneous, characterized by elevated levels of hydrogen peroxide (H_2O_2) and GSH, which promote tumor progression [63, 64]. In chemodynamic therapy (CDT), intracellular H_2O_2 is converted to hydroxyl ions, inducing DNA damage and apoptosis in the TME, thereby inhibiting tumor growth [65]. To enhance the effectiveness of CDT, RNA interference (RNAi) is coupled with it to form a brain-targeting biomimetic nanomedicine. This nanomedicine consists of a metastatic tumor cell membrane cloaked around the therapeutic payload, obtained from highly metastatic melanoma cells. Within the inner layer of the nanomedicine, molecules are present that induce membrane rupture, releasing the therapeutic siRNA complexed with polyethyleneimine xanthate (PEX). PEX destabilizes copper homeostasis, depletes GSH, and generates hydroxyl radicals (•OH). Combined with siRNA-mediated silencing of the Bcl-2 gene, this initiates multiple cascade reactions leading to tumor cell apoptosis and effective treatment of brain tumors [66].

A self-assembling fluorescent virus-like particle/RNAi nanocomplex (VLP/RNAi) is synthesized in *Escherichia coli* and further modified with cell-penetrating peptide and apolipoprotein E peptide, facilitating its crossing of the BBB. Once inside the brain, it inhibits the DNA repair mechanism and works in conjunction with TMZ to enhance chemotherapeutic effects. TMZ downregulates the hepatocyte growth factor receptor, and when combined with the VLP/RNAi system, it synergistically inhibits brain tumor growth [67]. This innovative approach holds promise for improving the efficacy of brain tumor treatment by targeting DNA repair mechanisms and enhancing the effects of chemotherapy.

Peptide nanofibers (PNFs) have emerged as effective carriers for delivering siRNA to brain tumor cells. These PNFs can be internalized by cells through endocytosis. When loaded with siRNA targeting polo-like kinase 1 (PLK1), PNFs reduce PLK1 mRNA expression and induce cell death in glioblastoma-derived stem cells. In mouse models of glioblastoma, intratumoral administration of PNF:siPLK1 significantly prolongs the lifespan of tumor-bearing mice [68]. These findings underscore the importance of nanomedicine-based RNA interference as a promising approach for brain tumor treatment [69].

Spherical nucleic acids (SNAs) represent a novel class of nanomaterials consisting of a central NP core surrounded by siRNA or miRNA molecules radiating outward. This unique structure shields the nucleic acids from degradation by nucleases and facilitates their entry into cells. When administered systemically, siRNA- or miRNA-conjugated SNAs have demonstrated the ability to penetrate both the BBB and the blood–tumor barrier. In preclinical studies using in vitro models and patient-derived xenograft murine models, these SNAs have shown promising results in inhibiting the growth of gliomas [70]. This highlights the potential of SNAs as effective vehicles for delivering nucleic acid-based therapies to treat brain tumors.

8.5 MONOCLONAL ANTIBODIES

mAbs of various origins, including murine, chimeric, humanized, and fully human, have been approved by regulatory agencies such as the FDA, European Medicines Agency, and other national authorities for treating a wide range of diseases. Examples of these approved mAbs include Rituximab for lymphoma [71], anti-PSMA/CD3 bispecific antibody [72], and CD19 × CD3 [73]. When encapsulated within polymer nanocapsules, Rituximab demonstrates sustained release properties, leading to up to a tenfold increase in mAb levels in the brain compared to naked mAb. Furthermore, functionalizing these nanocapsules with the CXCL13 ligand, which binds to the CXCR5 receptor overexpressed on B-cell lymphoma, can enhance targeting specificity. This treatment modality shows promise for treating cancers with brain metastasis [74].

Another innovative approach involves utilizing the chimeric anti-tenascin mAb 81C6 (ch81C6) functionalized with α-particle-emitting 211At for targeted radiotherapy of GBM. This strategy demonstrates low neurotoxicity and provides proof of concept for targeted radiotherapy to treat malignant brain tumors [75]. These advancements underscore the potential of mAbs and targeted radiotherapy in improving outcomes for patients with brain tumors.

Liposomes encapsulating mAbs can serve as effective carriers for drug delivery. In one study, long-circulating liposomes modified with the monoclonal anticancer antibody 2C5 (2C5-DoxLCL) were loaded with doxorubicin for the treatment of U-87 MG human brain tumors. These modified liposomes exhibited enhanced accumulation within U-87 MG tumors in mouse models, leading to superior therapeutic efficacy compared to control formulations. Furthermore, treatment with 2C5-DoxLCL resulted in a reduction in tumor size and a doubling in lifespan, highlighting its potential as a promising treatment approach for brain tumors (Figure 8.3) [76].

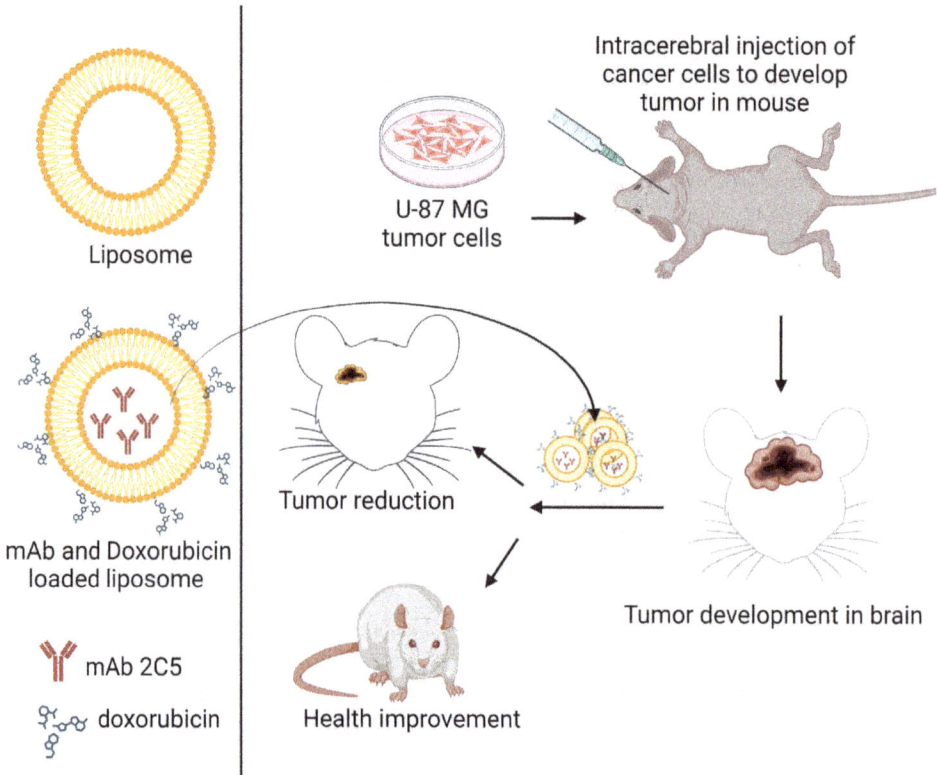

FIGURE 8.3 Treatment of brain tumors in mice using mAb and doxorubicin-loaded liposomes. This figure illustrates the process of treating brain tumors in mice using liposomes loaded with monoclonal antibodies (mAb 2C5) and doxorubicin. The steps include the development of brain tumors in mice through intracerebral injection of U-87 MG tumor cells, followed by the administration of liposomes containing mAb 2C5 and doxorubicin. The treatment leads to tumor reduction and health improvement in the mice.

8.6 DRUG REPURPOSING

Drug repurposing, a method where previously approved drugs are used for new indications, has gained significant attention across various medical fields. This approach offers several advantages, including saving time on new drug development, efficacy testing, and regulatory approval processes, as well as leveraging the known pharmacokinetic and safety profiles of existing drugs [77].

In the context of breast cancer, computational analysis has been utilized to identify FDA-approved drugs that may have efficacy against the disease. Similarly, this approach holds promise for targeting brain and other tumors. Tan et al. (2018) reviewed both psychiatric and non-psychiatric compounds for their potential repurposing [78].

One example is metformin, a drug indicated for type II diabetes. Metformin exerts its effects by decreasing hepatic glucose production and activating glucose uptake, thereby repressing mTOR activity. Given the enhanced mTOR activity observed in GBM, inhibiting mTOR activity is considered a key therapeutic approach against GBM. Treatment with metformin has been shown to sensitize GBM cells to TMZ administration, resulting in reduced stemness and viability in GBM tumorspheres [79–81].

Another repurposed drug is propentofylline (PPF), a phosphodiesterase inhibitor known for reducing inflammation in microglial cells associated with vascular dementia and Alzheimer's disease. PPF exhibits good brain accumulation ability and targets the TROY receptor, which is overexpressed in glioma cells. By decreasing TROY expression, PPF has been demonstrated to reduce the invasion and survival of GBM cells [82, 83]. These examples underscore the potential of drug repurposing as a strategy to identify novel treatments for brain tumors.

Pimozide, primarily used to treat antipsychotic disorders and Tourette's syndrome, schizophrenia, and bipolar disorder, has shown promise in the treatment of GBM. Pimozide treatment has been found to reduce the migration and survival of GBM cells and exhibits synergistic effects when combined with TMZ [84].

Similarly, chlorpromazine, another antipsychotic medication prescribed for similar mental health conditions, has shown the potential to inhibit glioma growth when combined with nitrosourea [85]. Studies have demonstrated that chlorpromazine treatment leads to inhibition of the PI3K/AKT/mTOR signaling pathway in C6 glioma cells [86]. Even in chemoresistant patient-derived glioma stem cells, chlorpromazine has shown effectiveness, although its targets are non-canonical [87]. These findings highlight the potential repurposing of antipsychotic drugs for the treatment of glioblastoma, offering new avenues for therapeutic interventions against this aggressive brain tumor.

In GBM cells, elevated levels of ALDH1 enzymatic activity have been observed. Disulfiram, a medication primarily used to treat chronic alcoholism, has been found to inhibit ALDH1 activity in these cells. By doing so, disulfiram disrupts cellular energetics and impairs the migration and viability of brain tumor cells [88].

Moreover, disulfiram has been shown to reduce the expression of the DNA repair enzyme O6-methylguanine-DNA methyltransferase (MGMT). This reduction sensitizes MGMT-treated cells to alkylating agents, which induce DNA damage and trigger DNA damage-mediated apoptosis [89]. These findings suggest that disulfiram holds potential as a therapeutic agent for GBM by targeting ALDH1 activity and sensitizing tumor cells to DNA-damaging agents.

Chloroquine, originally used as an antimalarial drug, has gained attention as an anti-tumor immune modulator. Its anticancer effects are multifaceted, involving both autophagy-dependent and autophagy-independent mechanisms [90–92]. Furthermore, chloroquine-mediated lysosomal dysfunction has been shown to enhance its anticancer activity [93]. These findings highlight the diverse mechanisms through which chloroquine exerts its anticancer effects, making it a promising candidate for cancer therapy.

Tf-CRM107, a conjugate of transferrin and a modified version of diphtheria toxin, selectively targets transferrin receptors overexpressed on tumor cells. Chloroquine has been found to reduce the toxicity of diphtheria toxin and Tf-CRM107. In a study evaluating its impact on U251 glioma-bearing nude mice, chloroquine treatment had minimal effect on the anti-tumor activity of Tf-CRM107, while effectively protecting normal vasculature. This suggests that chloroquine may enhance the therapeutic window of Tf-CRM107 during brain tumor therapy [94]. Additionally, when chloroquine is administered in combination with TMZ, synergistic anti-tumor effects have been observed in gliomas. This synergy is attributed to the modulation of autophagy [95]. These findings underscore the potential of chloroquine as an adjunctive therapy to enhance the efficacy of existing treatments for brain tumors.

8.7 CONCLUSION

This chapter has explored various innovative approaches in brain cancer therapy, highlighting advancements in nanomedicine, drug repurposing, and RNA interference technologies aimed at overcoming the complexities of malignant brain tumors. Nanomedicines have shown potential for delivering therapeutics effectively across the BBB, while drug repurposing offers a quicker transition to clinical applications by utilizing existing medications known for their anticancer effects. Additionally, the integration of mAbs into treatment regimens enhances targeted delivery, improving drug efficacy and reducing side effects. Furthermore, RNA interference technologies offer promising methods for silencing genes that support tumor growth and resistance, presenting a potential revolution in treatment paradigms. Collectively, these advancements suggest a hopeful future for brain cancer treatment, focusing on efficacy and safety to improve patient outcomes. The continued development and integration of these innovative strategies are critical in addressing the challenges posed by brain tumors.

REFERENCES

[1] Leece, R.; Xu, J.; Ostrom, Q.T.; Chen, Y.; Kruchko, C.; Barnholtz-Sloan, J.S. Global Incidence of Malignant Brain and Other Central Nervous System Tumors by Histology, 2003–2007. *Neuro Oncol*, 2017, *19*, 1553–1564.

[2] Decraene, B.; Vanmechelen, M.; Clement, P.; Daisne, J.F.; Vanden Bempt, I.; Sciot, R.; Garg, A.D.; Agostinis, P.; De Smet, F.; De Vleeschouwer, S. Cellular and Molecular Features Related to Exceptional Therapy Response and Extreme Long-Term Survival in Glioblastoma. *Cancer Medicine*, 2023, *12*(10), 11107–11126.

[3] Madani, F.; Esnaashari, S.S.; Webster, T.J.; Khosravani, M.; Adabi, M. Polymeric Nanoparticles for Drug Delivery in Glioblastoma: State of the Art and Future Perspectives. *J Control Release*, 2022, *349*, 649–661.

[4] Liu, G.; Qiu, Y.; Zhang, P.; Chen, Z.; Chen, S.; Huang, W.; Wang, B.; Yu, X.; Guo, D. Immunogenic Cell Death Enhances Immunotherapy of Diffuse Intrinsic Pontine Glioma: From Preclinical to Clinical Studies. *Pharmaceutics*, 2022, *14*, 1762.

[5] Huang, Z.; Dewanjee, S.; Chakraborty, P.; Jha, N.K.; Dey, A.; Gangopadhyay, M.; Chen, X.-Y.; Wang, J.; Jha, S.K. CAR T Cells: Engineered Immune Cells to Treat Brain Cancers and Beyond. *Mol Cancer*, 2023, *22*, 22.

[6] Simone, V.; Rizzo, D.; Cocciolo, A.; Caroleo, A.M.; Carai, A.; Mastronuzzi, A.; Tornesello, A. Infantile Brain Tumors: A Review of Literature and Future Perspectives. *Diagnostics (Basel)*, 2021, *11*, 670.

[7] DeCordova, S.; Shastri, A.; Tsolaki, A.G.; Yasmin, H.; Klein, L.; Singh, S.K.; Kishore, U. Molecular Heterogeneity and Immunosuppressive Microenvironment in Glioblastoma. *Front Immunol*, 2020, *11*, 1402.

[8] Upton, D.H.; Ung, C.; George, S.M.; Tsoli, M.; Kavallaris, M.; Ziegler, D.S. Challenges and Opportunities to Penetrate the Blood-Brain Barrier for Brain Cancer Therapy. *Theranostics*, 2022, *12*, 4734–4752.

[9] Engle, K.; Kumar, G. Cancer Multidrug-Resistance Reversal by ABCB1 Inhibition: A Recent Update. *Eur J Med Chem*, 2022, *239*, 114542.

[10] Shakir, S.; Wiheeb, A.; Khalaf, Z.; Othman, M.R. Improved Carbon Dioxide Capture by Nanofluids Containing Inorganic Nanoparticles and Binding Organic Liquid. *Periódico Tchê Química*, 2020, *17*, 688–705.

[11] A. Hassan, B.; Kadhim Lawi, Z.; Banoon, S. Detecting the Activity of Silver Nanoparticles, Pseudomonas Fluorescens and Bacillus Circulans on Inhibition of Aspergillus Niger Growth Isolated From Moldy Orange Fruits. *Periodico Tche Quimica*, 2020, *17*, 678–690.

[12] Farhan, A.; Jadah, A. Study of the Kinetics of Thermal Degradation of Unsaturated Polyester and Polyester/Silica Nanoparticles Composites by TGA and DSC Analysis Techniques. *Periodico Tche Quimica*, 2020, *17*, 437.

[13] Neto, J.; Pontes, R.; Rocha, P.; Marinho, F.; Silva, M. Experimental Analysis of a Solar Heat System Using Hybrid Silver Nanofluid and Titanium Dioxide. *Periodico Tche Quimica*, 2020, *17*, 448.

[14] Aldujaili, N.; Banoon, S. Antibacterial Characterization of Titanium Nanoparticles Nanosynthesized by Streptococcus Thermophilus. *Periodico Tche Quimica*, 2020, *17*, 311–320.

[15] Ioni, Y. Nanoparticles of Noble Metals on the Surface of Graphene Flakes. *Periódico Tchê Química*, 2020, *17*, 1199–1211.

[16] Rocha, P.A.; Santos, R.F.; Lima, R.J.; da Silva, M.E. A Review on Nanofluids: Preparation Methods and Applications. *Periodico Tche Quimica*, 2019, *16*(31), 365–380.

[17] Wiwatchaitawee, K.; Ebeid, K.; Quarterman, J.C.; Naguib, Y.; Ali, M.Y.; Oliva, C.; Griguer, C.; Salem, A.K. Surface Modification of Nanoparticles Enhances Drug Delivery to the Brain and Improves Survival in a Glioblastoma Multiforme Murine Model. *Bioconjug Chem*, 2022, *33*, 1957–1972.

[18] Jain, D.; Bajaj, A.; Athawale, R.; Shrikhande, S.; Goel, P.N.; Nikam, Y.; Gude, R.; Patil, S.; Prashant Raut, P. Surface-Coated PLA Nanoparticles Loaded with Temozolomide for Improved Brain Deposition and Potential Treatment of Gliomas: Development, Characterization and in vivo Studies. *Drug Deliv*, 2016, *23*, 999–1016.

[19] Zhang, Y.; Fu, X.; Jia, J.; Wikerholmen, T.; Xi, K.; Kong, Y.; Wang, J.; Chen, H.; Ma, Y.; Li, Z.; Wang, C.; Qi, Q.; Thorsen, F.; Wang, J.; Cui, J.; Li, X.; Ni, S. Glioblastoma Therapy Using Codelivery of

Cisplatin and Glutathione Peroxidase Targeting SiRNA from Iron Oxide Nanoparticles. *ACS Appl Mater Interfaces*, 2020, *12*, 43408–43421.

[20] Martínez-Sánchez, M.V.; Periago, A.; Legaz, I.; Gimeno, L.; Mrowiec, A.; Montes-Barqueros, N.R.; Campillo, J.A.; Bolarin, J.M.; Bernardo, M.V.; López-Álvarez, M.R.; González, C.; García-Garay, M.C.; Muro, M.; Cabañas-Perianes, V.; Fuster, J.L.; García-Alonso, A.M.; Moraleda, J.M.; Álvarez-Lopez, M.R.; Minguela, A. Overexpression of KIR Inhibitory Ligands (HLA-I) Determines That Immunosurveillance of Myeloma Depends on Diverse and Strong NK Cell Licensing. *Oncoimmunology*, 2015, *5*, e1093721.

[21] Baiula, M.; Cirillo, M.; Martelli, G.; Giraldi, V.; Gasparini, E.; Anelli, A.C.; Spampinato, S.M.; Giacomini, D. Selective Integrin Ligands Promote Cell Internalization of the Antineoplastic Agent Fluorouracil. *ACS Pharmacol Transl Sci*, 2021, *4*, 1528–1542.

[22] Toporkiewicz, M.; Meissner, J.; Matusewicz, L.; Czogalla, A.; Sikorski, A.F. Toward a Magic or Imaginary Bullet? Ligands for Drug Targeting to Cancer Cells: Principles, Hopes, and Challenges. *Int J Nanomedicine*, 2015, *10*, 1399–1414.

[23] Guo, J.; Schlich, M.; Cryan, J.F.; O'Driscoll, C.M. Targeted Drug Delivery via Folate Receptors for the Treatment of Brain Cancer: Can the Promise Deliver? *J Pharm Sci*, 2017, *106*, 3413–3420.

[24] Choudhury, H.; Pandey, M.; Chin, P.X.; Phang, Y.L.; Cheah, J.Y.; Ooi, S.C.; Mak, K.-K.; Pichika, M.R.; Kesharwani, P.; Hussain, Z.; Gorain, B. Transferrin Receptors-Targeting Nanocarriers for Efficient Targeted Delivery and Transcytosis of Drugs into the Brain Tumors: A Review of Recent Advancements and Emerging Trends. *Drug Deliv Transl Res*, 2018, *8*, 1545–1563.

[25] Muñoz, M.; Coveñas, R. Neurokinin-1 Receptor Antagonists as Antitumor Drugs in Gastrointestinal Cancer: A New Approach. *Saudi J Gastroenterol*, 2016, *22*, 260–268.

[26] Byeon, H.J.; Thao, L.Q.; Lee, S.; Min, S.Y.; Lee, E.S.; Shin, B.S.; Choi, H.-G.; Youn, Y.S. Doxorubicin-Loaded Nanoparticles Consisted of Cationic- and Mannose-Modified-Albumins for Dual-Targeting in Brain Tumors. *J Control Release*, 2016, *225*, 301–313.

[27] Cui, Y.; Xu, Q.; Chow, P.K.-H.; Wang, D.; Wang, C.-H. Transferrin-Conjugated Magnetic Silica PLGA Nanoparticles Loaded with Doxorubicin and Paclitaxel for Brain Glioma Treatment. *Biomaterials*, 2013, *34*, 8511–8520.

[28] Achrol, A.S.; Rennert, R.C.; Anders, C.; Soffietti, R.; Ahluwalia, M.S.; Nayak, L.; Peters, S.; Arvold, N.D.; Harsh, G.R.; Steeg, P.S.; Chang, S.D. Brain Metastases. *Nat Rev Dis Primers*, 2019, *5*, 5.

[29] Ni, J.; Miao, T.; Su, M.; Khan, N.U.; Ju, X.; Chen, H.; Liu, F.; Han, L. PSMA-Targeted Nanoparticles for Specific Penetration of Blood-Brain Tumor Barrier and Combined Therapy of Brain Metastases. *J Control Release*, 2021, *329*, 934–947.

[30] Zhang, W.; Sigdel, G.; Mintz, K.J.; Seven, E.S.; Zhou, Y.; Wang, C.; Leblanc, R.M. Carbon Dots: A Future Blood-Brain Barrier Penetrating Nanomedicine and Drug Nanocarrier. *Int J Nanomedicine*, 2021, *16*, 5003–5016.

[31] Hettiarachchi, S.D.; Graham, R.M.; Mintz, K.J.; Zhou, Y.; Vanni, S.; Peng, Z.; Leblanc, R.M. Triple Conjugated Carbon Dots as a Nano-Drug Delivery Model for Glioblastoma Brain Tumors. *Nanoscale*, 2019, *11*, 6192–6205.

[32] Liyanage, P.Y.; Zhou, Y.; Al-Youbi, A.O.; Bashammakh, A.S.; El-Shahawi, M.S.; Vanni, S.; Graham, R.M.; Leblanc, R.M. Pediatric Glioblastoma Target-Specific Efficient Delivery of Gemcitabine across the Blood-Brain Barrier via Carbon Nitride Dots. *Nanoscale*, 2020, *12*, 7927–7938.

[33] Kim, H.S.; Lee, D.Y. Nanomedicine in Clinical Photodynamic Therapy for the Treatment of Brain Tumors. *Biomedicines*, 2022, *10*, 96.

[34] Xu, X.; Chong, Y.; Liu, X.; Fu, H.; Yu, C.; Huang, J.; Zhang, Z. Multifunctional Nanotheranostic Gold Nanocages for Photoacoustic Imaging Guided Radio/Photodynamic/Photothermal Synergistic Therapy. *Acta Biomater*, 2019, *84*, 328–338.

[35] Xu, W.; Qian, J.; Hou, G.; Wang, Y.; Wang, J.; Sun, T.; Ji, L.; Suo, A.; Yao, Y. A Dual-Targeted Hyaluronic Acid-Gold Nanorod Platform with Triple-Stimuli Responsiveness for Photodynamic/Photothermal Therapy of Breast Cancer. *Acta Biomater*, 2019, *83*, 400–413.

[36] Shih, C.-Y.; Huang, W.-L.; Chiang, I.-T.; Su, W.-C.; Teng, H. Biocompatible Hole Scavenger-Assisted Graphene Oxide Dots for Photodynamic Cancer Therapy. *Nanoscale*, 2021, *13*, 8431–8441.

[37] Mangalath, S.; Saneesh Babu, P.S.; Nair, R.R.; Manu, P.M.; Krishna, S.; Nair, S.A.; Joseph, J. Graphene Quantum Dots Decorated with Boron Dipyrromethene Dye Derivatives for Photodynamic Therapy. *ACS Appl Nano Mater.*, 2021, *4*, 4162–4171.

[38] Roeinfard, M.; Zahedifar, M.; Darroudi, M.; Khorsand Zak, A.; Sadeghi, E. Preparation and Characterization of Selenium-Decorated Graphene Quantum Dots with High Afterglow for Application in Photodynamic Therapy. *Luminescence*, 2020, *35*, 891–896.

[39] Rzhepakovsky, I.V.; Areshidze, D.A.; Avanesyan, S.S.; Grimm, W.D.; Filatova, N.V.; Kalinin, A.V.; Kochergin, S.G.; Kozlova, M.A.; Kurchenko, V.P.; Sizonenko, M.N.; Terentiev, A.A.; Timchenko, L.D.; Trigub, M.M.; Nagdalian, A.A.; Piskov, S.I. Phytochemical Characterization, Antioxidant Activity, and Cytotoxicity of Methanolic Leaf Extract of Chlorophytum Comosum (Green Type) (Thunb.) Jacq. *Molecules*, 2022, *27*, 762.

[40] Rodriguez-Almaraz, J.E.; Butowski, N. Therapeutic and Supportive Effects of Cannabinoids in Patients with Brain Tumors (CBD Oil and Cannabis). *Curr Treat Options Oncol*, 2023, *24*, 30–44.

[41] Velasco, G.; Sánchez, C.; Guzmán, M. Towards the Use of Cannabinoids as Antitumour Agents. *Nat Rev Cancer*, 2012, *12*, 436–444.

[42] Peeri, H.; Shalev, N.; Vinayaka, A.C.; Nizar, R.; Kazimirsky, G.; Namdar, D.; Anil, S.M.; Belausov, E.; Brodie, C.; Koltai, H. Specific Compositions of Cannabis Sativa Compounds Have Cytotoxic Activity and Inhibit Motility and Colony Formation of Human Glioblastoma Cells In Vitro. *Cancers (Basel)*, 2021, *13*, 1720.

[43] Duzan, A.; Reinken, D.; McGomery, T.L.; Ferencz, N.M.; Plummer, J.M.; Basti, M.M. Endocannabinoids Are Potential Inhibitors of Glioblastoma Multiforme Proliferation. *J Integr Med*, 2023, *21*, 120–129.

[44] Sredni, S.T.; Huang, C.-C.; Suzuki, M.; Pundy, T.; Chou, P.; Tomita, T. Spontaneous Involution of Pediatric Low-Grade Gliomas: High Expression of Cannabinoid Receptor 1 (CNR1) at the Time of Diagnosis May Indicate Involvement of the Endocannabinoid System. *Childs Nerv Syst*, 2016, *32*, 2061–2067.

[45] Chen, C.-H.; Lin, M.-L.; Ong, P.-L.; Yang, J.-T. Novel Multiple Apoptotic Mechanism of Shikonin in Human Glioma Cells. *Ann Surg Oncol*, 2012, *19*, 3097–3106.

[46] Zhao, Q.; Kretschmer, N.; Bauer, R.; Efferth, T. Shikonin and Its Derivatives Inhibit the Epidermal Growth Factor Receptor Signaling and Synergistically Kill Glioblastoma Cells in Combination with Erlotinib. *Int J Cancer*, 2015, *137*, 1446–1456.

[47] Matias, D.; Balça-Silva, J.; Dubois, L.G.; Pontes, B.; Ferrer, V.P.; Rosário, L.; do Carmo, A.; Echevarria-Lima, J.; Sarmento-Ribeiro, A.B.; Lopes, M.C.; Moura-Neto, V. Dual Treatment with Shikonin and Temozolomide Reduces Glioblastoma Tumor Growth, Migration and Glial-to-Mesenchymal Transition. *Cell Oncol (Dordr)*, 2017, *40*, 247–261.

[48] Li, H.; Tong, Y.; Bai, L.; Ye, L.; Zhong, L.; Duan, X.; Zhu, Y. Lactoferrin Functionalized PEG-PLGA Nanoparticles of Shikonin for Brain Targeting Therapy of Glioma. *Int J Biol Macromol*, 2018, *107*, 204–211.

[49] Ding, D.; Wei, S.; Song, Y.; Li, L.; Du, G.; Zhan, H.; Cao, Y. Osthole Exhibits Anti-Cancer Property in Rat Glioma Cells through Inhibiting PI3K/Akt and MAPK Signaling Pathways. *Cell Physiol Biochem*, 2013, *32*, 1751–1760.

[50] Tsai, C.-F.; Yeh, W.-L.; Chen, J.-H.; Lin, C.; Huang, S.-S.; Lu, D.-Y. Osthole Suppresses the Migratory Ability of Human Glioblastoma Multiforme Cells via Inhibition of Focal Adhesion Kinase-Mediated Matrix Metalloproteinase-13 Expression. *Int J Mol Sci*, 2014, *15*, 3889–3903.

[51] Ngulde, S.I.; Sandabe, U.K.; Abounader, R.; Dawson, T.K.; Zhang, Y.; Iliya, I.; Hussaini, I.M. Ethanol Extract of Securidaca Longipedunculata Induces Apoptosis in Brain Tumor (U87) Cells. *Biomed Res Int*, 2019, *2019*, 9826590.

[52] Filippi-Chiela, E.C.; Villodre, E.S.; Zamin, L.L.; Lenz, G. Autophagy Interplay with Apoptosis and Cell Cycle Regulation in the Growth Inhibiting Effect of Resveratrol in Glioma Cells. *PLoS One*, 2011, *6*, e20849.

[53] Ngameni, B.; Touaibia, M.; Patnam, R.; Belkaid, A.; Sonna, P.; Ngadjui, B.T.; Annabi, B.; Roy, R. Inhibition of MMP-2 Secretion from Brain Tumor Cells Suggests Chemopreventive Properties of a Furanocoumarin Glycoside and of Chalcones Isolated from the Twigs of Dorstenia Turbinata. *Phytochemistry*, 2006, *67*, 2573–2579.

[54] Naveed, M.; Hejazi, V.; Abbas, M.; Kamboh, A.A.; Khan, G.J.; Shumzaid, M.; Ahmad, F.; Babazadeh, D.; FangFang, X.; Modarresi-Ghazani, F.; WenHua, L.; XiaoHui, Z. Chlorogenic Acid (CGA): A Pharmacological Review and Call for Further Research. *Biomed Pharmacother*, 2018, *97*, 67–74.

[55] Saeed, M.; Naveed, M.; Arif, M.; Kakar, M.U.; Manzoor, R.; Abd El-Hack, M.E.; Alagawany, M.; Tiwari, R.; Khandia, R.; Munjal, A.; Karthik, K.; Dhama, K.; Iqbal, H.M.N.; Dadar, M.; Sun, C. Green Tea (Camellia Sinensis) and l-Theanine: Medicinal Values and Beneficial Applications in Humans-A Comprehensive Review. *Biomed Pharmacother*, 2017, *95*, 1260–1275.

[56] Wang, X.; Lu, H.; Urvalek, A.M.; Li, T.; Yu, L.; Lamar, J.; DiPersio, C.M.; Feustel, P.J.; Zhao, J. KLF8 Promotes Human Breast Cancer Cell Invasion and Metastasis by Transcriptional Activation of MMP9. *Oncogene*, 2011, *30*, 1901–1911.

[57] Xu, C.; Hu, D.; Zhu, Q. EEF1A2 Promotes Cell Migration, Invasion and Metastasis in Pancreatic Cancer by Upregulating MMP-9 Expression through Akt Activation. *Clin Exp Metastasis*, 2013, *30*, 933–944.

[58] Pardridge, W.M. Intravenous, Non-Viral RNAi Gene Therapy of Brain Cancer. *Expert Opin Biol Ther*, 2004, *4*, 1103–1113.

[59] Ulbrich, K.; Knobloch, T.; Kreuter, J. Targeting the Insulin Receptor: Nanoparticles for Drug Delivery across the Blood-Brain Barrier (BBB). *J Drug Target*, 2011, *19*, 125–132.

[60] Zhang, Y.; Zhu, C.; Pardridge, W.M. Antisense Gene Therapy of Brain Cancer with an Artificial Virus Gene Delivery System. *Mol Ther*, 2002, *6*, 67–72.

[61] Podlecki, D.A.; Smith, R.M.; Kao, M.; Tsai, P.; Huecksteadt, T.; Brandenburg, D.; Lasher, R.S.; Jarett, L.; Olefsky, J.M. Nuclear Translocation of the Insulin Receptor. A Possible Mediator of Insulin's Long Term Effects. *J Biol Chem*, 1987, *262*, 3362–3368.

[62] Yu, D.; Khan, O.F.; Suvà, M.L.; Dong, B.; Panek, W.K.; Xiao, T.; Wu, M.; Han, Y.; Ahmed, A.U.; Balyasnikova, I.V.; Zhang, H.F.; Sun, C.; Langer, R.; Anderson, D.G.; Lesniak, M.S. Multiplexed RNAi Therapy against Brain Tumor-Initiating Cells via Lipopolymeric Nanoparticle Infusion Delays Glioblastoma Progression. *Proc Natl Acad Sci U S A*, 2017, *114*, E6147–E6156.

[63] Lim, M.; Xia, Y.; Bettegowda, C.; Weller, M. Current State of Immunotherapy for Glioblastoma. *Nat Rev Clin Oncol*, 2018, *15*, 422–442.

[64] Seo, S.; Nah, S.-Y.; Lee, K.; Choi, N.; Kim, H.N. Triculture Model of In Vitro BBB and Its Application to Study BBB-Associated Chemosensitivity and Drug Delivery in Glioblastoma. *Adv Funct Mater*, 2022, *32*, 2106860.

[65] Lin, L.-S.; Song, J.; Song, L.; Ke, K.; Liu, Y.; Zhou, Z.; Shen, Z.; Li, J.; Yang, Z.; Tang, W.; Niu, G.; Yang, H.-H.; Chen, X. Simultaneous Fenton-like Ion Delivery and Glutathione Depletion by MnO_2-Based Nanoagent to Enhance Chemodynamic Therapy. *Angew Chem Int Ed Engl*, 2018, *57*, 4902–4906.

[66] Zhang, D.; Sun, Y.; Wang, S.; Zou, Y.; Zheng, M.; Shi, B. Brain-Targeting Metastatic Tumor Cell Membrane Cloaked Biomimetic Nanomedicines Mediate Potent Chemodynamic and RNAi Combinational Therapy of Glioblastoma. *Adv Funct Mater*, 2022, *32*, 2209239.

[67] Pang, H.-H.; Huang, C.-Y.; Chou, Y.-W.; Lin, C.-J.; Zhou, Z.-L.; Shiue, Y.-L.; Wei, K.-C.; Yang, H.-W. Bioengineering Fluorescent Virus-like Particle/RNAi Nanocomplexes Act Synergistically with Temozolomide to Eradicate Brain Tumors. *Nanoscale*, 2019, *11*, 8102–8109.

[68] Liu, Y.; Zou, Y.; Feng, C.; Lee, A.; Yin, J.; Chung, R.; Park, J.B.; Rizos, H.; Tao, W.; Zheng, M.; Farokhzad, O.C. Charge Conversional Biomimetic Nanocomplexes as a Multifunctional Platform for Boosting Orthotopic Glioblastoma RNAi Therapy. *Nano Letters*, 2020, *20*(3), 1637–1646.

[69] Mazza, M.; Ahmad, H.; Hadjidemetriou, M.; Agliardi, G.; Pathmanaban, O.N.; King, A.T.; Bigger, B.W.; Vranic, S.; Kostarelos, K. Hampering Brain Tumor Proliferation and Migration Using Peptide Nanofiber:SiPLK1/MMP2 Complexes. *Nanomedicine (Lond)*, 2019, *14*, 3127–3142.

[70] Tommasini-Ghelfi, S.; Lee, A.; Mirkin, C.A.; Stegh, A.H. Synthesis, Physicochemical, and Biological Evaluation of Spherical Nucleic Acids for RNAi-Based Therapy in Glioblastoma. *Methods Mol Biol*, 2019, *1974*, 371–391.

[71] Sato, Y.; Watanabe, S.; Kodama, T.; Goto, M.; Shimosato, Y. Stainability of Lung Cancer Cells with Leu-7 and OKT-9 Monoclonal Antibodies. *Jpn J Clin Oncol*, 1985, *15*, 537–544.

[72] Leconet, W.; Liu, H.; Guo, M.; Le Lamer-Déchamps, S.; Molinier, C.; Kim, S.; Vrlinic, T.; Oster, M.; Liu, F.; Navarro, V.; Batra, J.S.; Noriega, A.L.; Grizot, S.; Bander, N.H. Anti-PSMA/CD3 Bispecific Antibody Delivery and Antitumor Activity Using a Polymeric Depot Formulation. *Mol Cancer Ther*, 2018, *17*, 1927–1940.

[73] Löffler, A.; Kufer, P.; Lutterbüse, R.; Zettl, F.; Daniel, P.T.; Schwenkenbecher, J.M.; Riethmüller, G.; Dörken, B.; Bargou, R.C. A Recombinant Bispecific Single-Chain Antibody, CD19 x CD3, Induces Rapid and High Lymphoma-Directed Cytotoxicity by Unstimulated T Lymphocytes. *Blood*, 2000, *95*, 2098–2103.

[74] Wen, J.; Wu, D.; Qin, M.; Liu, C.; Wang, L.; Xu, D.; Vinters, H.V.; Liu, Y.; Kranz, E.; Guan, X.; Sun, G.; Sun, X.; Lee, Y.; Martinez-Maza, O.; Widney, D.; Lu, Y.; Chen, I.S.Y.; Kamata, M. Sustained Delivery and Molecular Targeting of a Therapeutic Monoclonal Antibody to Metastases in the Central Nervous System of Mice. *Nat Biomed Eng*, 2019, *3*, 706–716.

[75] Zalutsky, M.R.; Reardon, D.A.; Akabani, G.; Coleman, R.E.; Friedman, A.H.; Friedman, H.S.; McLendon, R.E.; Wong, T.Z.; Bigner, D.D. Clinical Experience with α-Particle–Emitting 211At: Treatment of Recurrent Brain Tumor Patients with 211At-Labeled Chimeric Antitenascin Monoclonal Antibody 81C6. *J Nucl Med*, 2008, *49*, 30–38.

[76] Gupta, B.; Torchilin, V.P. Monoclonal Antibody 2C5-Modified Doxorubicin-Loaded Liposomes with Significantly Enhanced Therapeutic Activity against Intracranial Human Brain U-87 MG Tumor Xenografts in Nude Mice. *Cancer Immunol Immunother*, 2007, *56*, 1215–1223.

[77] Nowak-Sliwinska, P.; Scapozza, L.; Ruiz i Altaba, A. Drug Repurposing in Oncology: Compounds, Pathways, Phenotypes and Computational Approaches for Colorectal Cancer. *Biochim Biophys Acta Rev Cancer*, 2019, *1871*, 434–454.

[78] Tan, S.K.; Jermakowicz, A.; Mookhtiar, A.K.; Nemeroff, C.B.; Schürer, S.C.; Ayad, N.G. Drug Repositioning in Glioblastoma: A Pathway Perspective. *Front Pharmacol*, 2018, *9*, 218.

[79] Inoki, K.; Zhu, T.; Guan, K.-L. TSC2 Mediates Cellular Energy Response to Control Cell Growth and Survival. *Cell*, 2003, *115*, 577–590.

[80] Yamada, D.; Hoshii, T.; Tanaka, S.; Hegazy, A.M.; Kobayashi, M.; Tadokoro, Y.; Ohta, K.; Ueno, M.; Ali, M.A.E.; Hirao, A. Loss of Tsc1 Accelerates Malignant Gliomagenesis When Combined with Oncogenic Signals. *J Biochem*, 2014, *155*, 227–233.

[81] Iliff, J.J.; Lee, H.; Yu, M.; Feng, T.; Logan, J.; Nedergaard, M.; Benveniste, H. Brain-Wide Pathway for Waste Clearance Captured by Contrast-Enhanced MRI. *J Clin Invest*, 2013, *123*, 1299–1309.

[82] Jacobs, V.L.; Liu, Y.; De Leo, J.A. Propentofylline Targets TROY, a Novel Microglial Signaling Pathway. *PLoS One*, 2012, *7*, e37955.

[83] Paulino, V.M.; Yang, Z.; Kloss, J.; Ennis, M.J.; Armstrong, B.A.; Loftus, J.C.; Tran, N.L. TROY (TNFRSF19) Is Overexpressed in Advanced Glial Tumors and Promotes Glioblastoma Cell Invasion via Pyk2-Rac1 Signaling. *Mol Cancer Res*, 2010, *8*, 1558–1567.

[84] Harder, B.G.; Blomquist, M.R.; Wang, J.; Kim, A.J.; Woodworth, G.F.; Winkles, J.A.; Loftus, J.C.; Tran, N.L. Developments in Blood-Brain Barrier Penetrance and Drug Repurposing for Improved Treatment of Glioblastoma. *Front Oncol*, 2018, *8*, 462.

[85] Aas, A.T.; Brun, A.; Pero, R.W.; Salford, L.G. Chlorpromazine in Combination with Nitrosourea Inhibits Experimental Glioma Growth. *Br J Neurosurg*, 1994, *8*, 187–192.

[86] Shin, S.Y.; Lee, K.S.; Choi, Y.-K.; Lim, H.J.; Lee, H.G.; Lim, Y.; Lee, Y.H. The Antipsychotic Agent Chlorpromazine Induces Autophagic Cell Death by Inhibiting the Akt/MTOR Pathway in Human U-87MG Glioma Cells. *Carcinogenesis*, 2013, *34*, 2080–2089.

[87] Oliva, C.R.; Zhang, W.; Langford, C.; Suto, M.J.; Griguer, C.E. Repositioning Chlorpromazine for Treating Chemoresistant Glioma through the Inhibition of Cytochrome c Oxidase Bearing the COX4-1 Regulatory Subunit. *Oncotarget*, 2017, *8*, 37568–37583.

[88] Mashimo, T.; Pichumani, K.; Vemireddy, V.; Hatanpaa, K.J.; Singh, D.K.; Sirasanagandla, S.; Nannepaga, S.; Piccirillo, S.G.; Kovacs, Z.; Foong, C.; Huang, Z.; Barnett, S.; Mickey, B.E.; DeBerardinis, R.J.; Tu, B.P.; Maher, E.A.; Bachoo, R.M. Acetate Is a Bioenergetic Substrate for Human Glioblastoma and Brain Metastases. *Cell*, 2014, *159*, 1603–1614.

[89] Paranjpe, A.; Zhang, R.; Ali-Osman, F.; Bobustuc, G.C.; Srivenugopal, K.S. Disulfiram Is a Direct and Potent Inhibitor of Human O6-Methylguanine-DNA Methyltransferase (MGMT) in Brain Tumor Cells and Mouse Brain and Markedly Increases the Alkylating DNA Damage. *Carcinogenesis*, 2014, *35*, 692–702.

[90] Chen, D.; Xie, J.; Fiskesund, R.; Dong, W.; Liang, X.; Lv, J.; Jin, X.; Liu, J.; Mo, S.; Zhang, T.; Cheng, F.; Zhou, Y.; Zhang, H.; Tang, K.; Ma, J.; Liu, Y.; Huang, B. Chloroquine Modulates Antitumor Immune Response by Resetting Tumor-Associated Macrophages toward M1 Phenotype. *Nat Commun*, 2018, *9*, 873.

[91] Kimura, T.; Takabatake, Y.; Takahashi, A.; Isaka, Y. Chloroquine in Cancer Therapy: A Double-Edged Sword of Autophagy. *Cancer Res*, 2013, *73*, 3–7.

[92] Maes, H.; Kuchnio, A.; Carmeliet, P.; Agostinis, P. Chloroquine Anticancer Activity Is Mediated by Autophagy-Independent Effects on the Tumor Vasculature. *Mol Cell Oncol*, 2016, *3*, e970097.

[93] Harhaji-Trajkovic, L.; Arsikin, K.; Kravic-Stevovic, T.; Petricevic, S.; Tovilovic, G.; Pantovic, A.; Zogovic, N.; Ristic, B.; Janjetovic, K.; Bumbasirevic, V.; Trajkovic, V. Chloroquine-Mediated Lysosomal Dysfunction Enhances the Anticancer Effect of Nutrient Deprivation. *Pharm Res*, 2012, *29*, 2249–2263.

[94] Hagihara, N.; Walbridge, S.; Olson, A.W.; Oldfield, E.H.; Youle, R.J. Vascular Protection by Chloroquine during Brain Tumor Therapy with Tf-CRM107. *Cancer Res*, 2000, *60*, 230–234.

[95] Lee, S.W.; Kim, H.-K.; Lee, N.-H.; Yi, H.-Y.; Kim, H.-S.; Hong, S.H.; Hong, Y.-K.; Joe, Y.A. The Synergistic Effect of Combination Temozolomide and Chloroquine Treatment Is Dependent on Autophagy Formation and P53 Status in Glioma Cells. *Cancer Lett*, 2015, *360*, 195–204.

9 Green Nanomedicine for Targeted Brain Tumor Therapy

Yawen Ma[1]*, Rekha Khandia[2], and Pankaj Gurjar[3]
[1]Department of Neurosurgery, Shengjing Hospital of China Medical University, Shenyang, Liaoning, China
[2]Department of Biochemistry and Genetics, Barkatullah University, Bhopal MP, India
[3]Department of Science and Engineering, Novel Global Community Educational Foundation, Hebersham, NSW Australia
*Corresponding author

9.1 INTRODUCTION

Despite significant advancements in nanotechnology for treating and diagnosing brain cancer over the past two decades, several challenges persist, including issues with biocompatibility, selectivity, and efficacy [1]. One major challenge involves the size of nanoparticles (NPs), which affects their targeting and distribution within brain tumors. Due to the hyper-vascularization and compromised lymphatic drainage systems in brain tumors, NPs smaller than 100 nm can passively accumulate within the tumor tissue [2]. However, these smaller NPs are prone to opsonization, limiting their effectiveness [3]. To overcome this limitation, researchers have focused on modifying the surface of NPs to prevent recognition and elimination by the reticuloendothelial system. This is achieved through the functionalization of NPs with various coatings and functional groups [4], which improves their targetability and reduces opsonization. The surface functionality of NPs plays a crucial role in their interaction with biological molecules. By altering the surface chemistry, scientists have been able to enhance the properties of NPs and improve their targetability. For instance, coating NPs with hydrophilic polymers makes them resistant to protein adsorption, thereby protecting them from immune cell recognition and prolonging their circulation half-life. This ultimately increases their chances of reaching the brain and improves their bioavailability, leading to enhanced treatment efficacy [1].

Functionalizing NPs often requires additional synthesis reactions and the use of substances that may pose toxicity risks to both human health and the environment. To address these concerns, researchers have introduced the concept of green nanotechnology, which involves utilizing biologically derived molecules as natural raw materials for NP synthesis, conjugation, and functionalization [5]. This approach is gaining traction, particularly in the field of cancer treatment, where green nanomedicines are being increasingly utilized. Despite its potential, there remains a notable scarcity of examples in the literature showcasing the application of green nanotechnology in this context. This chapter aims to address this gap by presenting the available information on the green synthesis of nanostructures and its implications in the treatment of brain tumors. By leveraging green synthesis methods, researchers can develop efficient and tailored solutions for managing brain tumors in a sustainable manner.

9.2 BACTERIA-DERIVED NANOTECHNOLOGIES

Bacteria offer a diverse range of biochemical properties ideal for constructing nanomaterials. They have been utilized in the bioremediation of heavy metals owing to their capability to reduce metal

DOI: 10.1201/9781003519706-9

ions. Through enzymatic processes, bacteria acquire metal ions from their surroundings and undergo biological processes either intracellularly or extracellularly. Numerous microbial species have been harnessed for the production of mineral nanostructures [6]. However, the main challenge and focus of research lie in controlling the shape and size of these nanostructures for their application in medical science [7].

Iron NPs, such as iron oxide (Fe3O4) NPs, are frequently utilized in targeting brain tumors. While these NPs are relatively easy to produce, they often require coating with polymers to reduce cytotoxicity. Despite the growing interest in using ferro/ferrimagnetic NPs for tumor theranostics, a significant challenge remains in achieving targeted internalization. Magnetosomes, which are produced by magnetotactic bacteria, offer a potential solution [8, 9]. These magnetosomes exhibit uniform size and morphology when the magnetotactic bacteria are cultivated under optimal conditions, with a standard size typically ranging between 45 and 55 nm [10].

Magnetosomes exhibit numerous desirable characteristics. First, they are naturally arranged in a chain-like structure inside bacteria, and even after the bacteria are disrupted for harvesting, this spatial arrangement remains undisturbed. Consequently, magnetosomes are resistant to aggregation [11]. Second, these magnetosomes are enveloped by a lipoprotein layer, which imparts a negative charge, ensuring good dispersibility in water [12] and high biocompatibility [13]. Additionally, the presence of various chemical groups on the surface of magnetosomes facilitates easy functionalization [14].

The effectiveness of magnetosomes in MRI was initially demonstrated by Benoit et al. in 2009. They injected *Magnetospirillum magneticum* AMB-1 intratumorally and intravenously, observing their localization in mouse tumor xenografts and enhanced MRI contrast [15]. Consequently, magnetosomes produced by magnetotactic bacteria present a promising avenue for visualizing tumors in both preclinical and translational studies [14]. Furthermore, magnetosomes decorated with arginine–glycine–aspartic acid were prepared and injected via the tail vein. They were evaluated for enhanced permeability and retention into U87 tumoral cells, leading to an MRI contrast enhancement. Histopathological analysis confirmed the intratumoral accumulation of magnetosomes [16]. Iron oxide NPs can be synthesized in a mixture of oleylamine and oleic acid, yielding magnetic NPs with sizes ranging between 14 and 100 nm [17]. NPs coated with polyethylene glycol-block-polycaprolactone and functionalized with the glioma-targeting ligand lactoferrin, along with a fluorescent probe (Cy5.5), exhibited accumulation within glioma cells. This accumulation facilitated the delineation of the tumor margin in MRI scans, thereby proving valuable for guiding surgical glioma resection by providing clear delineation of tumor boundaries [18]. The administration of magnetosomes in intracranial U87-MG and subcutaneous GL-261 glioma tumors, combined with multiple alternating magnetic field applications, resulted in the complete disappearance of glioblastoma multiforme (GBM) with no observable brain damage. Additionally, the magnetosome treatment effectively prevented the recurrence of GBM tumors in mouse models [19]. This experiment offers compelling evidence of the therapeutic promise of biologically derived magnetosomes in eradicating infiltrating GBM cells.

9.3   MAMMALIAN CELL-DERIVED NANOTECHNOLOGIES

Various techniques have been developed to achieve effective drug delivery into tumor cells using NPs, but the size of the NPs remains a limiting factor. Scientists have made strides in synthesizing NPs inside cells using environmentally friendly approaches. Gold NPs have been formulated within various cell types including HEK-293 (human embryonic kidney), HeLa (human cervical cancer), SiHa (human cervical cancer), and SKNSH (human neuroblastoma) cells. The ability of cancer cells to facilitate the reduction process is crucial for synthesizing NPs, which is challenging with normal healthy cells [20]. Examples of cancer cell-mediated synthesis of NPs can be found in various literature sources [20–24]. For imaging cancer cells, an aqueous solution containing Fe^{2+} and Zn^{2+} ions is introduced into cancer cells. This leads to the synthesis of fluorescent ZnO nanoclusters and magnetic Fe_3O_4 nanoclusters, which can be utilized for imaging cancer cells through magnetic resonance imaging and computed tomography imaging [25]. NG108–15 neuroblastoma-glioma brain cancer cells have been observed to produce intracellular gold NPs (AuNPs) within 24 hours when a gold salt solution is added, with the generated fluorescence serving as proof of NP generation [26].

9.4   VIRUS-DERIVED/MIMETIC NANOSTRUCTURES

The surface proteins of viruses play a crucial role in facilitating their cellular uptake, which can be leveraged to deliver proteins or drugs into cells. Mimicking viral proteins can serve as a means to internalize desired molecules within cells, enhancing their therapeutic potential. The rabies virus is a neurotropic virus that causes hydrophobia and is 100% fatal in its advanced stages. The rabies virus glycoprotein (RVG) is responsible for transporting the virus to the central nervous system, crossing the blood–brain barrier (BBB), and facilitating viral entry into neurons. RVG acts as a ligand for the nicotinic acetylcholine receptor (AchR) expressed on neuronal cells [27]. Gold nanorods have shown improved cellular entry and delivery due to their larger transmembrane transport and effective hydrodynamic diameter. In a study by Changkyu et al. (2017), rabies virus-mimetic silica-coated gold nanorods (RVG-PEG-AuNRs@SiO2) were synthesized to target brain gliomas. These nanorods were functionalized with a 29-mer RVG peptide to enable the targeting of neural cells. Upon irradiation with an 808 nm laser, the rabies virus-mimetic NPs induced a hyperthermal effect, resulting in the reduction of brain tumors (Figure 9.1) [28].

Chitosan-dextran superparamagnetic nanoparticles (CS-DX-SPIONs) have demonstrated enhanced tumor imaging capabilities and enable targeted delivery of chemotherapeutic agents [29].

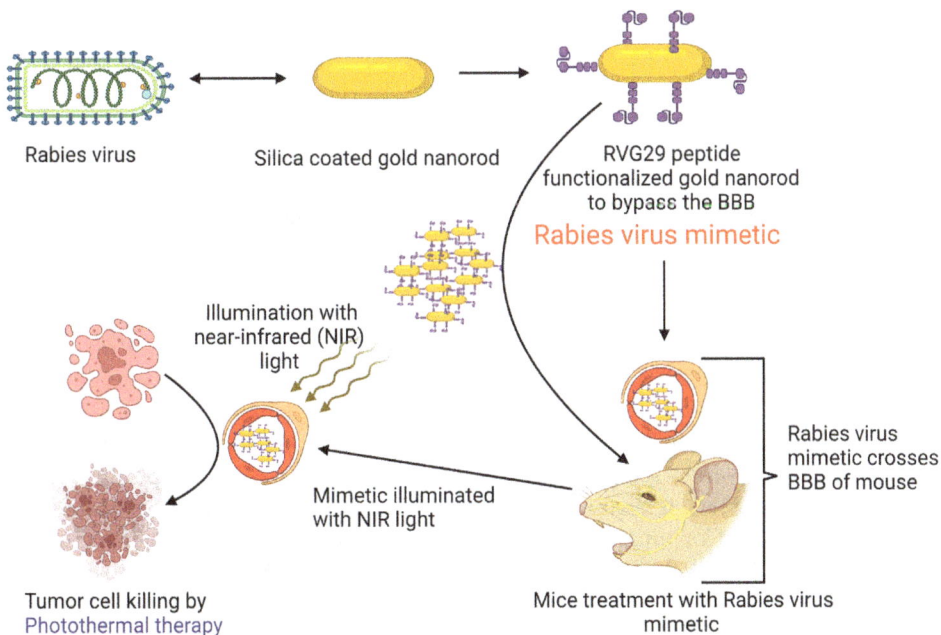

FIGURE 9.1   Nanoparticle-mediated photothermal therapy for cancer treatment utilizing a rabies virus-mimetic system. The schematic depicts the development and application of silica-coated gold nanorods functionalized with the RVG29 peptide. These nanorods mimic the rabies virus to bypass the blood–brain barrier (BBB) in mice. Initially, the silica-coated gold nanorods are synthesized and then functionalized with RVG29 peptides. Upon administration, these nanoparticles cross the BBB effectively, resembling the mechanism employed by the rabies virus. Tumor cells targeted by these nanoparticles are exposed to near-infrared (NIR) light, leading to their destruction via photothermal therapy, demonstrated in both in vitro and in vivo settings.

Mitoxantrone (MTO) is an effective chemotherapeutic drug for treating common and aggressive brain tumors. However, its efficacy is hindered by poor penetrability. To address this issue, MTO has been encapsulated within the cavity of cowpea mosaic virus (CPMV). This formulation, known as CPMV-MTO, has shown uptake by glioma cells and displayed anti-tumor effects [30].

9.5   NUCLEIC ACID-BASED NPS

NPs offer an effective means of delivering various types of nucleic acids, including small interfering RNA (siRNA), microRNA (miRNA), and messenger RNA (mRNA) [31]. By intelligently designing NPs, it is possible to overcome barriers such as heterogeneity and the BBB, thus paving the way for personalized and precision medicine [32]. Lipid nanoparticles (LNPs), typically spherical in shape, are commonly utilized for delivering nucleic acids as therapeutic agents. Comprising complex cationic lipids and negatively charged nucleic acids, LNPs facilitate endosomal escape [32]. Bioinspired lipoproteins (bLPs) have been engineered to remodel the tumor stromal microenvironment through photothermia, enhancing the access of other bLPs to tumor cells and thereby improving therapeutic efficacy [33]. PEGylation of NPs serves to protect their surface from aggregation, opsonization, and phagocytosis, thereby prolonging circulation and facilitating drug and gene delivery [34]. The immunosuppressive nature of the tumor microenvironment (TME), largely influenced by factors like tumor growth factor β (TGF-β), hampers the efficacy of chemotherapy. RNAi-based immunomodulation can modify the TME, enhancing anti-tumor effects. Dual systems, such as the emozolomide-siRNA system, have been developed to simultaneously induce tumor cell toxicity and siRNA-mediated gene silencing, resulting in prolonged survival in glioma tumor-bearing animal models. This system can be tailored to the specific tumor and target genes, operating synergistically [35]. Polymeric siRNA nanomedicines with improved physiological stability have been engineered to enable on-site delivery of siRNA in the presence of abundant reactive oxygen species. By functionalizing these nanomedicines with angiopep-2 peptide, efficient BBB crossing is facilitated. These nanomedicines effectively target polo-like kinase 1 (overexpressed in brain tumor cells) and vascular endothelial growth factor receptor-2, effectively suppressing GBM [36]. Such highly stabilized siRNA platforms provide robust and potent siRNA delivery for RNA interference therapy against brain tumors [37]. Nanocapsules containing angiopep-2-functionalized siRNA were synthesized by encapsulating siRNA within a capsule-like structure using positively charged acrylate guanidine and N, N'-bis(acryloyl) cystamine as a cross-linker. These nanocapsules efficiently crossed the blood–brain barrier (BBB), accumulated within GBM, and released siRNA at the target site [37]. In orthotopic U87MG xenografts, the nanocapsules inhibited tumor growth without apparent harmful effects and significantly enhanced mean survival [38]. A biomimetic NP platform based on red blood cell membrane (RBCm) has been developed, containing siRNA for effective silencing of GBM genes. This nanosystem, guided by angiopep-2, offers an immune reaction-free biomimetic platform for treating U87MG-luc human GBM tumor-bearing nude mice [39]. Chlototxin, derived from the Israeli yellow scorpion, is a peptide with 36 amino acids that specifically binds to GBM cells in a dose-dependent manner while sparing normal cells. Chlototxin also possesses anti-angiogenic activity [40]. Coupling chlototxin with a nanoscopic high-branching dendrimer, polyamidoamine (PAMAM), which binds specifically to receptors on GBM cells, has shown promising results. Cellular uptake of CTX-modified DNA dendrimers was demonstrated in C6 glioma cells, leading to higher survival times in glioma-bearing animals compared to the temozolomide-treated group [41]. The therapeutic efficacy of temozolomide (TMZ) is compromised by the DNA repair enzyme O6-methylguanine-DNA methyltransferase (MGMT). An iron oxide NP system has been developed for targeted siRNA delivery to suppress MGMT (Figure 9.2). Proof of concept has been demonstrated in GBM patient-derived xenografts, where sequential administration of NP and TMZ resulted in enhanced apoptosis of GBM stem cells and prolonged survival [42]. Another approach to silence MGMT genes is to methylate its promoter region, as a lack of methylation of the MGMT promoter is associated with TMZ resistance [43].

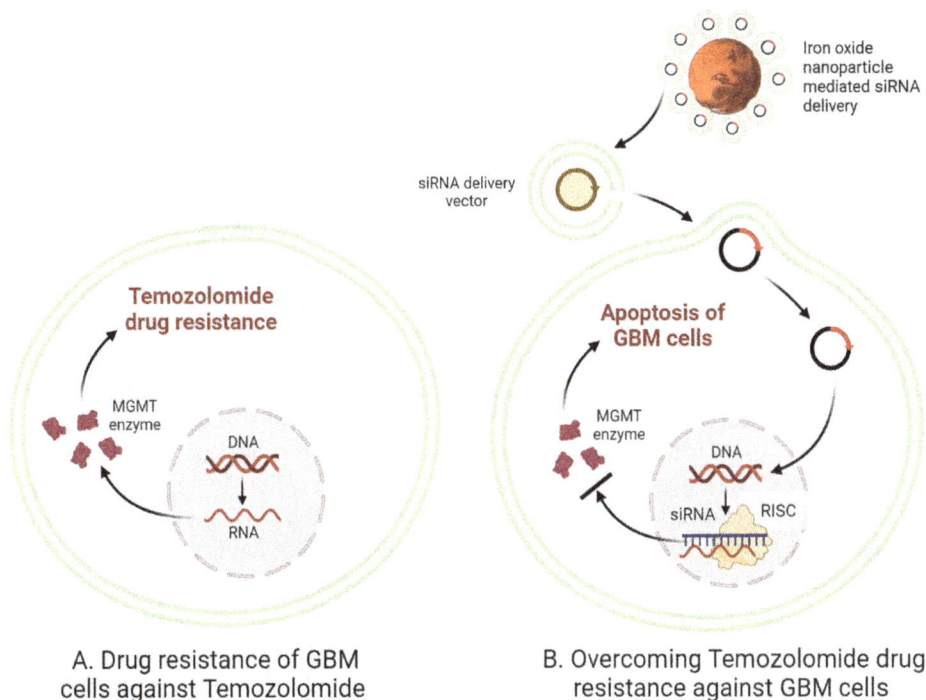

A. Drug resistance of GBM cells against Temozolomide

B. Overcoming Temozolomide drug resistance against GBM cells

FIGURE 9.2 Mechanisms of temozolomide resistance and siRNA-mediated overcoming of resistance in glioblastoma multiforme (GBM) cells. Panel A illustrates the cellular mechanism of temozolomide resistance in GBM cells, showing the role of the MGMT enzyme in repairing DNA damage induced by the drug, leading to resistance. Panel B depicts a strategy to overcome this resistance using iron oxide nanoparticles to deliver siRNA targeting the MGMT gene, thereby promoting apoptosis in resistant GBM cells.

Chlorotoxin-labeled NPs composed of an iron oxide NP core, chitosan, polyethylene glycol, and polyethyleneimine (PEI) were employed for targeted gene delivery to C6 glioma cells. Furthermore, intravenous injection of these NPs in mouse C6 xenografts demonstrated enhanced uptake, indicating their potential for targeted gene delivery in gliomas and other lethal tumors [44].

Co-delivery of STAT3siRNA with a small-molecule inhibitor of the JAK/STAT pathway using α5β1 integrin receptor-selective liposomes demonstrated effective internalization into GL261 cells. This approach significantly inhibited tumor growth in an animal model [45]. Moreover, neuropilin-1 (NRP-1) overexpression has been observed in U87 glioma cells. A peptide, RGERPPR, specifically binds to NRP-1 and can be utilized to target ligands for GBM cells. When this peptide is bound to branched PEI and plasmid DNA, it exhibits enhanced uptake in U87 glioma tumor cells, serving as an effective gene delivery vehicle for therapeutic DNA molecules [46].

A virus-mimetic nanogel has been developed, wherein therapeutic microRNA miR155, which regulates the phenotype of microglia and macrophages, is embedded into an erythrocyte membrane. This nanogel is further functionalized with M2pep peptides and HA2 peptides. The erythrocyte membrane serves to protect miR155 from degradation, while M2pep enhances targetability, and HA2 peptide, derived from the influenza virus, facilitates fusion between the erythrocyte membrane and endosomal membrane. Once in the tumor cytoplasm, miR155 is released from the nanogel through the action of ribonuclease H, exerting anti-tumor effects [47].

Spherical nucleic acid (SNA) NP conjugates consist of gold NPs and small interfering RNA duplexes. SNAs demonstrate efficient entry into and transformation of glial cells. In this context, the Bcl2Likc12 (Bcl2L12) oncoproteins, hyperactivated in glioma cells, are targeted. Intratumoral expression of these SNAs induces tumor cell apoptosis and leads to decreased tumor size.

9.6   CARBOHYDRATE-BASED NANOPLATFORMS

Carbohydrates are attractive for biomedical applications due to their biocompatibility and biological origin, existing as aldoses or ketoses. Chitosan, for instance, possesses free amino groups on its surface [48]. Being a natural alkaline polysaccharide, chitosan is resistant to enzymatic degradation. Hybrid chitosan-dextran superparamagnetic NPs (CS-DX-SPIONs) have been investigated for internalization into U87, C6 glioma, and HeLa cells. Upon intravenous administration, these NPs accumulated within GBM cells, providing high-contrast magnetic resonance imaging of tumor cells and enabling targeted drug delivery [29].

Chitosan NPs loaded with small peptides, such as the caspase inhibitor Z-DEVD-FMK, or large peptides like basic fibroblast growth factor, have demonstrated the ability to traverse the BBB. Functionalization of these peptides with transferrin enhances targetability and receptor-mediated transcytosis. Systemic administration of these formulations has been shown to provide neuroprotection [49].

Magnetic nanocarriers are valuable for imaging and drug delivery, particularly in the case of brain tumors. Chitosan NPs functionalized with transferrin ligand conjugated with Docetaxel have shown enhanced cellular uptake and cytotoxicity in C6 glioma cells. The targeted NPs exhibited significantly higher cytotoxicity and improved relative bioavailability of Docetaxel, demonstrating promising clinical efficacy with improved pharmacokinetics in brain tumor therapy [50].

In an innovative approach, polyamidoamine (PAMAM) dendrimers were conjugated with chitosan and temozolomide (TMZ) and tested both in vitro and in vivo against GBM. The synthesized conjugates underwent thorough characterization, including analysis by 1H NMR, FT-IR spectroscopy, evaluation of surface morphological parameters, and a hemolytic assay. Results revealed that the conjugate exhibited enhanced efficacy compared to TMZ alone against U-251 and T-98G glioma cell lines, along with a longer half-life. This novel strategy utilized chitosan-anchored dendrimers for TMZ delivery, highlighting its potential for improving GBM treatment [51].

Photodynamic therapy (PDT) combines photosensitizers (PS) that generate molecular oxygen upon exposure to specific wavelengths of light. In brain tumor research, PDT has garnered considerable interest. Gold nanoparticles (AuNPs) serve as excellent carriers for drugs and photothermal agents [52]. To enhance drug delivery, a gold nanocage structure has been developed for encapsulating small drug molecules [53]. However, AuNPs have low biocompatibility, necessitating coating with biomolecules [54].

In one approach, a photosensitizer entrapped within a gold nanocage is coated with a biocompatible polymer, glycol chitosan (GC). Additionally, an enzyme-responsive system is incorporated to ensure the controlled release of the encapsulated photosensitizer upon reaching the target site. Incorporating a cathepsin-B-responsive short peptide enhances phototoxicity against U87 brain tumor cells in vitro and reduces tumor growth in xenograft mouse models (Figure 9.3) [55].

Nanoformulations were developed using hyaluronic acid and chitosan hydrochloride (CSH), loaded with curcuminoid. These formulations demonstrated efficacy against glioma cells (C6) by efficiently crossing the BBB. This suggests their potential for further exploration in the treatment of brain malignancies [56].

9.7   PROTEIN-BASED NANOPLATFORMS

Superparamagnetic iron oxide nanoparticles find extensive applications in medical technology, but concerns regarding toxicity and off-target effects persist. To address these concerns, bovine serum albumin (BSA)-coated superparamagnetic iron oxide NPs have been engineered to target the internalization of glioma U251 cells. These BSA-conjugated NPs demonstrate undetectable toxicity and excellent biocompatibility. Moreover, when conjugated with fluorescein isothiocyanate (FITC), they enable visualization of tumor cells [57].

NPs utilizing cationic BSA have also been developed, showing strong permeability across the BBB. These engineered NPs were evaluated against HNGC1 tumors, exhibiting potent cytotoxicity and providing location information through FITC labeling [58].

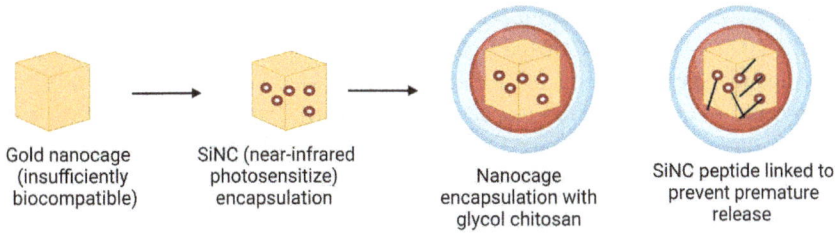

Generation of photosensitizer encapsulated gold nanocage for glioblastoma phototherapy

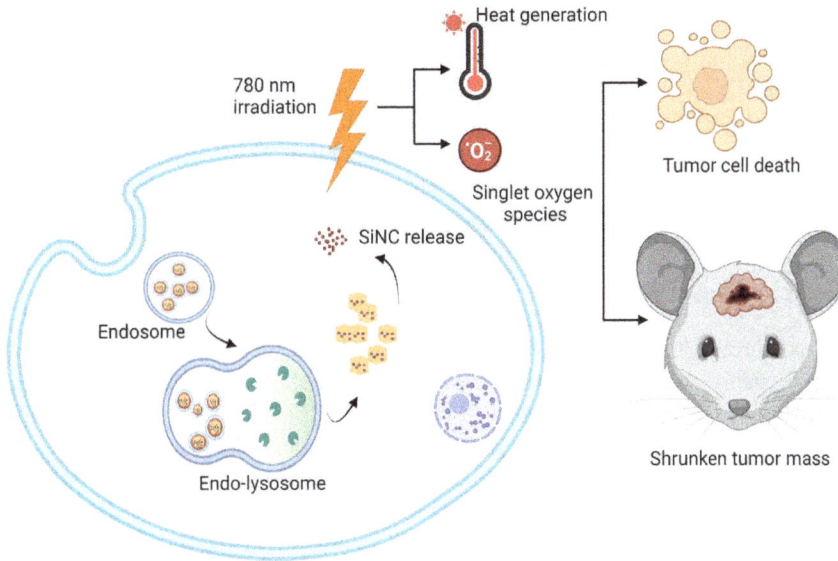

Effects of developed photosensitizer based module against cancer

FIGURE 9.3 Development and mechanism of action of a gold nanocage encapsulated photosensitizer for glioblastoma phototherapy. This illustration outlines the synthesis and therapeutic application of a photosensitizer-encapsulated gold nanocage for the treatment of glioblastoma. The process starts with a basic gold nanocage, which is modified by encapsulating it with a near-infrared photosensitizer (SiNC), and further stabilized with glycol chitosan to prevent premature release of the photosensitizer. The diagram also details the cellular uptake and therapeutic action, highlighting the induction of heat and singlet oxygen species upon 780 nm irradiation, leading to tumor cell death and shrinkage of tumor mass in a mouse model.

Additionally, albumin-templated Gd_2O_3 and MnO_2 NPs have been synthesized as MRI contrast-enhancing agents for imaging gliomas. MnO2@BSA NPs offer superior MRI contrast due to the gradual release of Mn ions in acidic brain tumor environments, making them promising candidates for enhanced imaging of gliomas [59].

Lactoferrin-functionalized and doxorubicin (DOX)-loaded BSA NPs have been developed and evaluated for their ability to penetrate the BBB and target glioma cells. These NPs were tested against primary brain capillary endothelial cells and glioma cells (C6), demonstrating good uptake and cytotoxicity specifically in brain tumor cells. Tissue distribution studies further revealed the accumulation of DOX in the brain, indicating the potential for targeted delivery of chemotherapeutic agents against brain tumors [60].

Overexpressed albumin-binding proteins such as SPARC and gp60 are found on various types of tumors, including gliomas, where they facilitate the trafficking of albumin to rapidly proliferating cancer cells. Leveraging this mechanism, low molecular weight protamine and a cell-penetrating peptide can guide the delivery of albumin NPs containing chemotherapeutic drugs like paclitaxel (PTX) and fenretinide (4-HPR) for targeted treatment of brain tumors. The hydrophobic nature of these chemotherapeutic agents promotes self-assembly of albumin NPs. 4-HPR, a vitamin A analog with pro-apoptotic properties against tumors, can be co-delivered with PTX to glioma cells via albumin NPs. The anti-tumor effect is attributed to several factors, including targeted delivery through the cell-penetrating peptide and selective binding and uptake facilitated by SPARC- and gp60-mediated biomimetic transport. Intracranial administration of the nanocomposite demonstrates improved anti-tumor effects [61].

Silk fibroin protein, known for its versatility in dyeing with a variety of colorants, has been utilized to encapsulate indocyanine green (ICG) for use as a therapeutic photothermal platform against GBM. These silk fibroin nanoparticles (SFNPs) exhibit a spherical morphology despite their fibrous nature and demonstrate good stability in physiological media. SFNPs enable sustained release of encapsulated ICG. Upon irradiation with an 808 nm laser beam for ten minutes, the temperature can be readily increased for photothermal therapy. The efficacy of SFNPs-ICG was evaluated in C6 glioma-bearing xenograft nude mice through in vivo imaging. ICG-loaded SFNPs effectively accumulated and induced cancer cell death upon irradiation in the near-infrared spectrum, suggesting their potential for efficient photothermal therapy of C6 glioma [62].

9.8 CONCLUSION

This chapter has highlighted the significant potential of green nanomedicines in the treatment of brain tumors, focusing on their biocompatibility, safety, and ability to precisely target tumor sites. Utilizing biologically derived materials, such as iron oxide NPs and magnetosomes, green nanomedicines enhance therapeutic efficacy while minimizing environmental impact. The various innovative approaches were discussed, including the use of virus-mimicking NPs and biocompatible coatings like GC, which improve the delivery and efficacy of therapeutics across the BBB. Techniques like photothermal therapy and targeted gene delivery have shown promising results in preclinical models, demonstrating the versatility and effectiveness of these nanotechnologies. As green nanomedicines continue to evolve, they hold great promise for advancing precision medicine, particularly for treating aggressive brain cancers like GBM. Further research and interdisciplinary collaboration are essential to overcome existing challenges and translate these technologies into clinical applications, ultimately leading to safer, more effective cancer treatments.

REFERENCES

[1] Nehra, M.; Uthappa, U.T.; Kumar, V.; Kumar, R.; Dixit, C.; Dilbaghi, N.; Mishra, Y.K.; Kumar, S.; Kaushik, A. Nanobiotechnology-Assisted Therapies to Manage Brain Cancer in Personalized Manner. *J Control Release*, 2021, *338*, 224–243.

[2] Lee, K.; David, A.E.; Zhang, J.; Shin, M.C.; Yang, V.C. Enhanced Accumulation of Theranostic Nanoparticles in Brain Tumor by External Magnetic Field Mediated in Situ Clustering of Magnetic Nanoparticles. *J Ind Eng Chem*, 2017, *54*, 389–397.

[3] Ruan, S.; Hu, C.; Tang, X.; Cun, X.; Xiao, W.; Shi, K.; He, Q.; Gao, H. Increased Gold Nanoparticle Retention in Brain Tumors by in Situ Enzyme-Induced Aggregation. *ACS Nano*, 2016, *10*, 10086–10098.

[4] Assa, F.; Jafarizadeh-Malmiri, H.; Ajamein, H.; Vaghari, H.; Anarjan, N.; Ahmadi, O.; Berenjian, A. Chitosan Magnetic Nanoparticles for Drug Delivery Systems. *Crit Rev Biotechnol*, 2017, *37*, 492–509.

[5] Nasrollahzadeh, M.; Sajjadi, M.; Sajadi, S.M.; Issaabadi, Z. Chapter 5 – Green Nanotechnology. In: *Interface Science and Technology*; Nasrollahzadeh, M.; Sajadi, S.M.; Sajjadi, M.; Issaabadi, Z.; Atarod, M., Eds.; An Introduction to Green Nanotechnology; Elsevier, 2019; Vol. 28, pp. 145–198.

[6] Zhang, X.; Yan, S.; Tyagi, R.D.; Surampalli, R.Y. Synthesis of Nanoparticles by Microorganisms and Their Application in Enhancing Microbiological Reaction Rates. *Chemosphere*, 2011, *82*, 489–494.

[7] Quax, T.E.F.; Claassens, N.J.; Söll, D.; van der Oost, J. Codon Bias as a Means to Fine-Tune Gene Expression. *Mol Cell*, 2015, *59*, 149–161.

[8] Bazylinski, D.A.; Garratt-Reed, A.J.; Frankel, R.B. Electron Microscopic Studies of Magnetosomes in Magnetotactic Bacteria. *Microsc Res Tech*, 1994, *27*, 389–401.

[9] Araujo, A.C.V.; Abreu, F.; Silva, K.T.; Bazylinski, D.A.; Lins, U. Magnetotactic Bacteria as Potential Sources of Bioproducts. *Mar Drugs*, 2015, *13*, 389–430.

[10] Taylor, A.P.; Barry, J.C. Magnetosomal Matrix: Ultrafine Structure May Template Biomineralization of Magnetosomes. *J Microsc*, 2004, *213*, 180–197.

[11] Chen, C.; Chen, L.; Yi, Y.; Chen, C.; Wu, L.-F.; Song, T. Killing of Staphylococcus Aureus via Magnetic Hyperthermia Mediated by Magnetotactic Bacteria. *Appl Environ Microbiol*, 2016, *82*, 2219–2226.

[12] Józefczak, A.; Leszczyński, B.; Skumiel, A.; Hornowski, T. A Comparison between Acoustic Properties and Heat Effects in Biogenic (Magnetosomes) and Abiotic Magnetite Nanoparticle Suspensions. *J Magn Magn Mater*, 2016, *407*, 92–100.

[13] Mickoleit, F.; Jörke, C.; Geimer, S.; Maier, D.S.; Müller, J.P.; Demut, J.; Gräfe, C.; Schüler, D.; Clement, J.H. Biocompatibility, Uptake and Subcellular Localization of Bacterial Magnetosomes in Mammalian Cells. *Nanoscale Adv*, 2021, *3*, 3799–3815.

[14] Vargas, G.; Cypriano, J.; Correa, T.; Leão, P.; Bazylinski, D.A.; Abreu, F. Applications of Magnetotactic Bacteria, Magnetosomes and Magnetosome Crystals in Biotechnology and Nanotechnology: Mini-Review. *Molecules*, 2018, *23*, 2438.

[15] Benoit, M.R.; Mayer, D.; Barak, Y.; Chen, I.Y.; Hu, W.; Cheng, Z.; Wang, S.X.; Spielman, D.M.; Gambhir, S.S.; Matin, A. Visualizing Implanted Tumors in Mice with Magnetic Resonance Imaging Using Magnetotactic Bacteria. *Clin Cancer Res*, 2009, *15*, 5170–5177.

[16] Boucher, M.; Geffroy, F.; Prévéral, S.; Bellanger, L.; Selingue, E.; Adryanczyk-Perrier, G.; Péan, M.; Lefèvre, C.T.; Pignol, D.; Ginet, N.; Mériaux, S. Genetically Tailored Magnetosomes Used as MRI Probe for Molecular Imaging of Brain Tumor. *Biomaterials*, 2017, *121*, 167–178.

[17] Xu, Z.; Shen, C.; Hou, Y.; Gao, H.; Sun, S. Oleylamine as Both Reducing Agent and Stabilizer in a Facile Synthesis of Magnetite Nanoparticles. https://pubs.acs.org/doi/pdf/10.1021/cm802978z (Accessed Apr 15, 2023).

[18] Zhou, Q.; Mu, K.; Jiang, L.; Xie, H.; Liu, W.; Li, Z.; Qi, H.; Liang, S.; Xu, H.; Zhu, Y.; Zhu, W.; Yang, X. Glioma-Targeting Micelles for Optical/Magnetic Resonance Dual-Mode Imaging. *Int J Nanomedicine*, 2015, *10*, 1805–1818.

[19] Alphandery, E.; Idbaih, A.; Adam, C.; Delattre, J.; Schmitt, C.; Guyot, F.; Chebbi, I. P04.10 Chains of Magnetosomes Induce Full Destruction of Intracranial U87-Luc and Subcutaneous GL-261 Glioma in Mice under the Application of an Alternating Magnetic Field. *Neuro Oncol*, 2018, *20*, iii280.

[20] Anshup; Venkataraman, J.S.; Subramaniam, C.; Kumar, R.R.; Priya, S.; Kumar, T.R.S.; Omkumar, R.V.; John, A.; Pradeep, T. Growth of Gold Nanoparticles in Human Cells. *Langmuir*, 2005, *21*, 11562–11567.

[21] Rehman, F.U.; Jiang, H.; Selke, M.; Wang, X. Mammalian Cells: A Unique Scaffold for in Situ Biosynthesis of Metallic Nanomaterials and Biomedical Applications. *J Mater Chem B*, 2018, *6*, 6501–6514.

[22] El-Said, W.A.; Cho, H.-Y.; Yea, C.-H.; Choi, J.-W. Synthesis of Metal Nanoparticles Inside Living Human Cells Based on the Intracellular Formation Process. *Adv Mater*, 2014, *26*, 910–918.

[23] Chen, D.; Zhao, C.; Ye, J.; Li, Q.; Liu, X.; Su, M.; Jiang, H.; Amatore, C.; Selke, M.; Wang, X. In Situ Biosynthesis of Fluorescent Platinum Nanoclusters: Toward Self-Bioimaging-Guided Cancer Theranostics. *ACS Appl Mater Interfaces*, 2015, *7*, 18163–18169.

[24] Cruz, D.M.; Mostafavi, E.; Vernet-Crua, A.; Barabadi, H.; Shah, V.; Cholula-Díaz, J.L.; Guisbiers, G.; Webster, T.J. Green Nanotechnology-Based Zinc Oxide (ZnO) Nanomaterials for Biomedical Applications: A Review. *J Phys Mater*, 2020, *3*, 034005.

[25] Du, T.; Zhao, C.; ur Rehman, F.; Lai, L.; Li, X.; Sun, Y.; Luo, S.; Jiang, H.; Selke, M.; Wang, X. Rapid and Multimodal in Vivo Bioimaging of Cancer Cells Through in Situ Biosynthesis of Zn&Fe Nanoclusters. *Nano Res*, 2017, *10*, 2626–2632.

[26] Gillan, M.; Zander, N. In-Vitro Synthesis of Gold Nanoclusters in Neurons. US Army Research Laboratory Aberdeen Proving Ground United States, 2016 Apr 1.

[27] Dani, J.A. Neuronal Nicotinic Acetylcholine Receptor Structure and Function and Response to Nicotine. *Int Rev Neurobiol*, 2015, *124*, 3–19.

[28] Lee, C.; Hwang, H.S.; Lee, S.; Kim, B.; Kim, J.O.; Oh, K.T.; Lee, E.S.; Choi, H.-G.; Youn, Y.S. Rabies Virus-Inspired Silica-Coated Gold Nanorods as a Photothermal Therapeutic Platform for Treating Brain Tumors. *Adv Mater*, 2017, *29*.

[29] Shevtsov, M.; Nikolaev, B.; Marchenko, Y.; Yakovleva, L.; Skvortsov, N.; Mazur, A.; Tolstoy, P.; Ryzhov, V.; Multhoff, G. Targeting Experimental Orthotopic Glioblastoma with Chitosan-Based Superparamagnetic Iron Oxide Nanoparticles (CS-DX-SPIONs). *Int J Nanomedicine*, 2018, *13*, 1471–1482.

[30] Lam, P.; Lin, R.; Steinmetz, N.F. Delivery of Mitoxantrone Using a Plant Virus-Based Nanoparticle for the Treatment of Glioblastomas. *J Mater Chem B*, 2018, *6*, 5888–5895.

[31] Riley, R.S.; June, C.H.; Langer, R.; Mitchell, M.J. Delivery Technologies for Cancer Immunotherapy. *Nat Rev Drug Discov*, 2019, *18*, 175–196.

[32] Mitchell, M.J.; Billingsley, M.M.; Haley, R.M.; Wechsler, M.E.; Peppas, N.A.; Langer, R. Engineering Precision Nanoparticles for Drug Delivery. *Nat Rev Drug Discov*, 2021, *20*, 101–124.

[33] Tan, T.; Hu, H.; Wang, H.; Li, J.; Wang, Z.; Wang, J.; Wang, S.; Zhang, Z.; Li, Y. Bioinspired Lipoproteins-Mediated Photothermia Remodels Tumor Stroma to Improve Cancer Cell Accessibility of Second Nanoparticles. *Nat Commun*, 2019, *10*, 3322.

[34] Suk, J.S.; Xu, Q.; Kim, N.; Hanes, J.; Ensign, L.M. PEGylation as a Strategy for Improving Nanoparticle-Based Drug and Gene Delivery. *Adv Drug Deliv Rev*, 2016, *99*, 28–51.

[35] Qiao, C.; Yang, J.; Shen, Q.; Liu, R.; Li, Y.; Shi, Y.; Chen, J.; Shen, Y.; Xiao, Z.; Weng, J.; Zhang, X. Traceable Nanoparticles with Dual Targeting and ROS Response for RNAi-Based Immunochemotherapy of Intracranial Glioblastoma Treatment. *Adv Mater*, 2018, *30*, e1705054.

[36] Lee, C.; Fotovati, A.; Triscott, J.; Chen, J.; Venugopal, C.; Singhal, A.; Dunham, C.; Kerr, J.M.; Verreault, M.; Yip, S.; Wakimoto, H.; Jones, C.; Jayanthan, A.; Narendran, A.; Singh, S.K.; Dunn, S.E. Polo-like Kinase 1 Inhibition Kills Glioblastoma Multiforme Brain Tumor Cells in Part through Loss of SOX2 and Delays Tumor Progression in Mice. *Stem Cells*, 2012, *30*, 1064–1075.

[37] Zheng, M.; Liu, Y.; Wang, Y.; Zhang, D.; Zou, Y.; Ruan, W.; Yin, J.; Tao, W.; Park, J.B.; Shi, B. ROS-Responsive Polymeric SiRNA Nanomedicine Stabilized by Triple Interactions for the Robust Glioblastoma Combinational RNAi Therapy. *Adv Mater*, 2019, *31*, e1903277.

[38] Zou, Y.; Sun, X.; Wang, Y.; Yan, C.; Liu, Y.; Li, J.; Zhang, D.; Zheng, M.; Chung, R.S.; Shi, B. Single SiRNA Nanocapsules for Effective SiRNA Brain Delivery and Glioblastoma Treatment. *Adv Mater*, 2020, *32*, e2000416.

[39] Liu, Y.; Zou, Y.; Feng, C.; Lee, A.; Yin, J.; Chung, R.; Park, J.B.; Rizos, H.; Tao, W.; Zheng, M.; Farokhzad, O.C.; Shi, B. Charge Conversional Biomimetic Nanocomplexes as a Multifunctional Platform for Boosting Orthotopic Glioblastoma RNAi Therapy. *Nano Lett*, 2020, *20*, 1637–1646.

[40] Cohen-Inbar, O.; Zaaroor, M. Glioblastoma Multiforme Targeted Therapy: The Chlorotoxin Story. *J Clin Neurosci*, 2016, *33*, 52–58.

[41] Huang, R.; Ke, W.; Han, L.; Li, J.; Liu, S.; Jiang, C. Targeted Delivery of Chlorotoxin-Modified DNA-Loaded Nanoparticles to Glioma via Intravenous Administration. *Biomaterials*, 2011, *32*, 2399–2406.

[42] Wang, K.; Kievit, F.M.; Chiarelli, P.A.; Stephen, Z.R.; Lin, G.; Silber, J.R.; Ellenbogen, R.G.; Zhang, M. SiRNA Nanoparticle Suppresses Drug-Resistant Gene and Prolongs Survival in an Orthotopic Glioblastoma Xenograft Mouse Model. *Adv Funct Mater*, 2021, *31*, 2007166.

[43] Kitange, G.J.; Carlson, B.L.; Schroeder, M.A.; Grogan, P.T.; Lamont, J.D.; Decker, P.A.; Wu, W.; James, C.D.; Sarkaria, J.N. Induction of MGMT Expression Is Associated with Temozolomide Resistance in Glioblastoma Xenografts. *Neuro Oncol*, 2009, *11*, 281–291.

[44] Kievit, F.M.; Veiseh, O.; Fang, C.; Bhattarai, N.; Lee, D.; Ellenbogen, R.G.; Zhang, M. Chlorotoxin Labeled Magnetic Nanovectors for Targeted Gene Delivery to Glioma. *ACS Nano*, 2010, *4*, 4587–4594.

[45] Vangala, V.; Nimmu, N.V.; Khalid, S.; Kuncha, M.; Sistla, R.; Banerjee, R.; Chaudhuri, A. Combating Glioblastoma by Codelivering the Small-Molecule Inhibitor of STAT3 and STAT3siRNA with A5β1 Integrin Receptor-Selective Liposomes. *Mol Pharm*, 2020, *17*, 1859–1874.

[46] Wang, J.; Lei, Y.; Xie, C.; Lu, W.; Yan, Z.; Gao, J.; Xie, Z.; Zhang, X.; Liu, M. Targeted Gene Delivery to Glioblastoma Using a C-End Rule RGERPPR Peptide-Functionalised Polyethylenimine Complex. *Int J Pharm*, 2013, *458*, 48–56.

[47] Gao, X.; Li, S.; Ding, F.; Liu, X.; Wu, Y.; Li, J.; Feng, J.; Zhu, X.; Zhang, C. A Virus-Mimicking Nucleic Acid Nanogel Reprograms Microglia and Macrophages for Glioblastoma Therapy. *Adv Mater*, 2021, *33*, e2006116.

[48] Das, S.N.; Wagenknecht, M.; Nareddy, P.K.; Bhuvanachandra, B.; Niddana, R.; Balamurugan, R.; Swamy, M.J.; Moerschbacher, B.M.; Podile, A.R. Amino Groups of Chitosan Are Crucial for Binding to a Family 32 Carbohydrate Binding Module of a Chitosanase from Paenibacillus Elgii. *J Biol Chem*, 2016, *291*, 18977–18990.

[49] Yemisci, M.; Caban, S.; Fernandez-Megia, E.; Capan, Y.; Couvreur, P.; Dalkara, T. Preparation and Characterization of Biocompatible Chitosan Nanoparticles for Targeted Brain Delivery of Peptides. *Methods Mol Biol*, 2018, *1727*, 443–454.

[50] Agrawal, P.; Singh, R.P.; Sonali, null; Kumari, L.; Sharma, G.; Koch, B.; Rajesh, C.V.; Mehata, A.K.; Singh, S.; Pandey, B.L.; Muthu, M.S. TPGS-Chitosan Cross-Linked Targeted Nanoparticles for Effective Brain Cancer Therapy. *Mater Sci Eng C Mater Biol Appl*, 2017, *74*, 167–176.

[51] Sharma, A.K.; Gupta, L.; Sahu, H.; Qayum, A.; Singh, S.K.; Nakhate, K.T.; Ajazuddin, null; Gupta, U. Chitosan Engineered PAMAM Dendrimers as Nanoconstructs for the Enhanced Anti-Cancer Potential and Improved In Vivo Brain Pharmacokinetics of Temozolomide. *Pharm Res*, 2018, *35*, 9.

[52] García Calavia, P.; Bruce, G.; Pérez-García, L.; Russell, D.A. Photosensitiser-Gold Nanoparticle Conjugates for Photodynamic Therapy of Cancer. *Photochem Photobiol Sci*, 2018, *17*, 1534–1552.

[53] Cobley, C.M.; Au, L.; Chen, J.; Xia, Y. Targeting Gold Nanocages to Cancer Cells for Photothermal Destruction and Drug Delivery. *Expert Opin Drug Deliv*, 2010, *7*, 577–587.

[54] Wang, Z.; Chen, Z.; Liu, Z.; Shi, P.; Dong, K.; Ju, E.; Ren, J.; Qu, X. A Multi-Stimuli Responsive Gold Nanocage-Hyaluronic Platform for Targeted Photothermal and Chemotherapy. *Biomaterials*, 2014, *35*, 9678–9688.

[55] Battogtokh, G.; Gotov, O.; Kang, J.H.; Hong, E.J.; Shim, M.S.; Shin, D.; Ko, Y.T. Glycol Chitosan-Coated near-Infrared Photosensitizer-Encapsulated Gold Nanocages for Glioblastoma Phototherapy. *Nanomedicine*, 2019, *18*, 315–325.

[56] Xu, Y.; Asghar, S.; Yang, L.; Li, H.; Wang, Z.; Ping, Q.; Xiao, Y. Lactoferrin-Coated Polysaccharide Nanoparticles Based on Chitosan Hydrochloride/Hyaluronic Acid/PEG for Treating Brain Glioma. *Carbohydr Polym*, 2017, *157*, 419–428.

[57] Wang, X.; Tu, M.; Tian, B.; Yi, Y.; Wei, Z.; Wei, F. Synthesis of Tumor-Targeted Folate Conjugated Fluorescent Magnetic Albumin Nanoparticles for Enhanced Intracellular Dual-Modal Imaging into Human Brain Tumor Cells. *Anal Biochem*, 2016, *512*, 8–17.

[58] Kesharwani, P.; Jain, A.; Jain, A.; Jain, A.K.; Garg, N.K.; Tekade, R.K.; Singh, T.R.R.; Iyer, A.K. Cationic Bovine Serum Albumin (CBA) Conjugated Poly Lactic-Co-Glycolic Acid (PLGA) Nanoparticles for Extended Delivery of Methotrexate into Brain Tumors. *RSC Adv.*, 2016, *6*, 89040–89050.

[59] Zhang, H.; Wang, T.; Zheng, Y.; Yan, C.; Gu, W.; Ye, L. Comparative Toxicity and Contrast Enhancing Assessments of Gd2O3@BSA and MnO2@BSA Nanoparticles for MR Imaging of Brain Glioma. *Biochem Biophys Res Commun*, 2018, *499*, 488–492.

[60] Su, Z.; Xing, L.; Chen, Y.; Xu, Y.; Yang, F.; Zhang, C.; Ping, Q.; Xiao, Y. Lactoferrin-Modified Poly(Ethylene Glycol)-Grafted BSA Nanoparticles as a Dual-Targeting Carrier for Treating Brain Gliomas. *Mol Pharm*, 2014, *11*, 1823–1834.

[61] Lin, T.; Zhao, P.; Jiang, Y.; Tang, Y.; Jin, H.; Pan, Z.; He, H.; Yang, V.C.; Huang, Y. Blood-Brain-Barrier-Penetrating Albumin Nanoparticles for Biomimetic Drug Delivery via Albumin-Binding Protein Pathways for Antiglioma Therapy. *ACS Nano*, 2016, *10*, 9999–10012.

[62] Xu, H.-L.; ZhuGe, D.-L.; Chen, P.-P.; Tong, M.-Q.; Lin, M.-T.; Jiang, X.; Zheng, Y.-W.; Chen, B.; Li, X.-K.; Zhao, Y.-Z. Silk Fibroin Nanoparticles Dyeing Indocyanine Green for Imaging-Guided Photo-Thermal Therapy of Glioblastoma. *Drug Deliv*, 2018, *25*, 364–375.

10 Drug Delivery Systems for Targeting Brain Tumors
Overcoming the Blood–Brain Barrier

Zirong Fan[1]*, Rekha Khandia[2], and Pankaj Gurjar[3]
[1]Department of Neurosurgery, Shengjing Hospital of China Medical University, Shenyang, Liaoning, China
[2]Department of Biochemistry and Genetics, Barkatullah University, Bhopal MP, India
[3]Department of Science and Engineering, Novel Global Community Educational Foundation, Hebersham, NSW Australia
*Corresponding author

10.1 INTRODUCTION

Cancer involves the uncontrolled growth of cells, with the potential to invade surrounding tissues—a process known as metastasis [1]. Brain tumors signify an unregulated proliferation of cells within brain tissue. When cancer originates elsewhere in the body and migrates to establish itself in the brain, it's termed a secondary brain tumor. The primary challenge in developing effective therapies for brain tumors lies in their heterogeneity and the presence of the blood–brain barrier (BBB) [2]. Targeting the BBB for clinical interventions requires efficient strategies. Nanoparticles (NPs) exhibit a propensity to accumulate in brain tissue, facilitated by enhanced permeability and retention effects resulting from the high permeability of tumor blood vessels. Moreover, NPs can be functionalized with active groups to enhance drug targeting [3]. Notably, significant progress has been achieved in the field of brain targeting in recent years.

The focus remained on mechanisms facilitating the invasion of the BBB with optimal bioavailability and sustained release. Targeted delivery strategies may involve exploring ligands of receptors that are overexpressed on brain tumor cells. Functionalization of these ligands, such as glucose, interleukin (IL6), or Angiopep-2, could enhance targeting efficiency [4–7]. Nanostructures, with their specific architecture, have the potential to penetrate cells, and the attachment of therapeutic and targeting moieties enables them to be highly specific to their targets.

Nanocages, capable of entrapping chemotherapeutic agents like DOX and paclitaxel, allow for sustained release and increased retention time, thereby maximizing the effectiveness of the drug. Hydrogels represent a recent advancement in cancer treatment. The utilization of thermos-reversible gelation polymers enables prolonged contact time and the delivery of various molecules, including DNA, peptides, nutraceuticals, and chemotherapeutic drugs.

10.2 TRANSPORT MECHANISMS

Drug molecules can be transported to the brain parenchyma via the trigeminal or olfactory nerve pathways [8]. In the extracellular pathways, drugs are absorbed through the nasal epithelium into the lamina propria and then transported through axons, a process referred to as bulk flux [8].

DOI: 10.1201/9781003519706-10

10.2.1 Nasal Transportation to the Brain

Nasal transport begins within the nasal cavity, where inhaled air interacts with the mucous-covered nasal surfaces [9]. The respiratory system also plays a role in drug delivery, benefiting from its highly vascularized structure to enhance drug absorption and facilitate increased drug transport to the brain [10]. Goblet cells within the mucus layer produce mucus, through which drugs penetrate to reach the epithelial surface. The viscosity of the mucus aids in preventing drug clearance [11]. The respiratory tract's high vascularization and expansive surface area enable efficient drug absorption, with therapeutic molecules being absorbed and transported through various neural pathways such as the trigeminal nerve, olfactory sense, or blood vessels [12].

10.2.2 Olfactory Neuronal Pathway

The olfactory neuronal pathway serves as a potent route for intranasal drug absorption [13], presenting an appealing, non-invasive method to deliver drugs while bypassing the BBB [14]. Drugs absorbed through this pathway are directly delivered to the brain via the mucosa, circumventing the BBB. Consequently, drug bioavailability in the brain is enhanced, with reduced risks of drug clearance and side effects. This route also allows for effective drug delivery by avoiding interaction with P-glycoprotein efflux proteins [15]. Despite its promising features, nasal drug delivery has limitations, including lower drug administration capacity and susceptibility to mucociliary and nasal enzymatic clearance [16]. To address these challenges, a polymer-based micelle system was developed, incorporating a cell-penetrating peptide to facilitate cellular invasion and loaded with the drug camptothecin. Intranasal administration of this micelle system demonstrated anti-glioma effects in C6 glioma cells. Moreover, the micelles exhibited intratumoral accumulation and prolonged survival in an animal model bearing C6 glioma orthotopic grafts [17]. These examples underscore the effectiveness of nasal drug delivery, offering a promising approach to circumvent issues associated with BBB permeability.

10.2.3 Trigeminal Nerve Pathways

When a drug molecule diffuses into the mucosa, it travels along the branches of the trigeminal nerve. During cellular uptake, various transport mechanisms come into play, including receptor-mediated, carrier-mediated, and transcytosis transport. Lipophilicity is a crucial parameter in drug design for targeting brain tissue [18]. A schematic illustration of a PEGylated Fe_3O_4 NP labeled with a fluorescent dye is provided in Figure 10.1, showcasing the developed module's dual fluorescent and magnetic resonance imaging capabilities.

10.2.4 Cerebrospinal Fluid Pathways

Cerebrospinal fluid (CSF) pathways extend from the subarachnoid space of the brain to the nasal lymphatic system [19]. Despite this understanding, detailed research reports on the CSF-lymphatic pathway from the nose to the brain are still limited, indicating the need for further investigation.

10.3 MODE OF TRANSPORTATION ACROSS THE BRAIN

Transport through the BBB primarily involves the following mechanisms.

10.3.1 Diffusion

This mechanism entails the passive movement of molecules across concentration gradients. It is particularly effective for small hydrophobic molecules, making it a critical strategy for the delivery of therapeutic molecules.

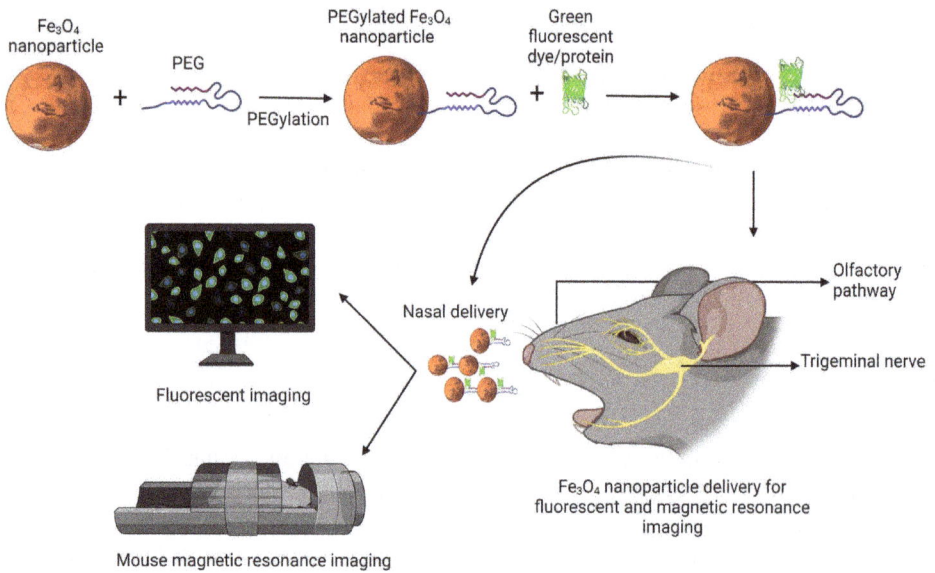

FIGURE 10.1 Delivery and imaging of Fe_3O_4 nanoparticles in mice. This figure demonstrates the preparation, administration, and tracking of PEGylated Fe_3O_4 nanoparticles conjugated with a green, fluorescent dye/protein in a mouse. The nanoparticles are nasally delivered and subsequently imaged using fluorescent and magnetic resonance imaging to monitor their pathway through the olfactory and trigeminal nerves.

10.3.2 PARACELLULAR TRANSPORT

Paracellular transport involves the movement of molecules through the intercellular space between adjacent endothelial cells. This route is typically limited to small water-soluble molecules. Gap junctions, including zonula adherens, tight junctions, and macula adherens, regulate paracellular transport by controlling the openings between cells, thereby facilitating the transport of molecules [20].

10.3.3 RECEPTOR-MEDIATED TRANSPORT

Efficient crossing of the BBB is essential for drug molecules intended for brain tumor therapy to exert their therapeutic effects. One strategy to achieve this is through receptor-mediated cellular transportation [21]. Receptors such as the transferrin receptor or insulin receptor facilitate the translocation of their ligands across the membrane. By conjugating drug molecules with these ligands, targeted therapy can be achieved [21].

In clinical practice, combinations of drugs like temozolomide and bevacizumab, along with a BCNU wafer for slow drug release, are commonly prescribed for glioma treatment [22]. Transferrin-coupled NPs functionalized with polyethylene glycol (PEG) can evade the reticuloendothelial system [23]. Insulin receptors play a crucial role in transporting their ligands to the brain parenchyma, making them potential targets for drug delivery. Humanized monoclonal antibodies are often utilized for delivery purposes in the brain [24].

Additionally, transferrin-coupled liposomes loaded with doxorubicin (DOX) have demonstrated efficient uptake by C6 glioma cells, exhibiting both stability and non-toxicity, along with targeted drug delivery capabilities [25]. Peptides such as the interleukin-6 receptor binding I6P7 peptide have been utilized for crossing the BBB and targeting gliomas as non-viral vectors for gene delivery to U87 glioma cells, demonstrating anti-tumor activity. Intravenous administration of the gene-bearing interleukin-6 ligand peptide has been shown to prolong survival in U87 glioma-bearing mice [26].

10.3.4 TRANSPORTERS

Specific transporters play crucial roles in binding to ligands and transporting them across the membrane. For instance, boron drugs are recognized as effective boron carriers that can cross the BBB via active transporters [27]. Various transporters, such as the glucose transporter [28], glutamate transporters [29], iron transporter [30], zinc transporter [31], and multidrug transport proteins [32], are expressed in brain tissue.

Elevated levels of ATP-binding cassette transporters are associated with drug resistance in gliomas [33]. Among these, ABCC4 and ABCC5 proteins are notably prevalent drug exporter proteins in gliomas, contributing significantly to drug resistance in brain cancer cells [34]. Understanding the roles and functions of these transporters is essential for developing effective strategies to overcome drug resistance and enhance drug delivery to brain tumors.

10.4   DIFFERENT NANOMATERIALS TO TARGET DRUG TO BRAIN TUMOR

During the golden age of pharmaceutical nanocarriers, nanostructures have emerged as invaluable tools for drug delivery. These nanostructures offer a combination of safety, efficacy, and scalability, making them suitable for large-scale industrial production and potential clinical use.

10.4.1 LIPID NPS

Lipid NPs and lipid nanocarriers are typically non-toxic, biocompatible, and straightforward to produce [35]. The concept of solid lipid NPs (SLN®) and nanostructured lipid carriers (NLC®) was introduced in 1990 [36]. NLCs differ from SLNs by incorporating small amounts of liquid lipids (oils) into their structure at room temperature, creating a structured matrix. These oils prevent drug release from the matrix, thus enhancing physicochemical stability [37].

Transcellular diffusion occurs for lipophilic molecules with a size of less than 450 kDa [38]. Larger molecules undergo endocytosis, primarily mediated through receptors. Apolipoprotein E (ApoE) possesses high-affinity receptors along the BBB, presenting an opportunity to exploit this property for trafficking drug molecules functionalized with receptor ligands [39].

10.4.2 DENDRIMERS

Dendrimers are highly branched, three-dimensional polymeric structures composed of a central core, iterative branching units, and multiple active surface groups. These features enable dendrimers to encapsulate a broad array of biomolecules, potentially reducing cytotoxicity and enhancing therapeutic efficacy [40]. Drugs can be loaded onto dendrimers either through covalent conjugation or electrostatic adsorption, allowing for precise drug delivery [41]. Owing to their size, which mimics that of biological molecules, dendrimers are particularly suited for drug delivery to the brain via nasal administration [42]. In one application, dendrimers were engineered with an internal Tamoxifen core and an external transferrin shell, facilitating efficient drug delivery across the BBB in glioma cells (Figure 10.2). This configuration also enables pH-triggered drug release at mildly acidic conditions, enhancing drug delivery through transferrin-mediated endocytosis and countering the effects of drug efflux transporters in C6 glioma cells [43]. Additionally, methotrexate (MTX)-loaded dendrimers, functionalized with d-glucosamine to act as ligands, proved more effective in crossing the BBB compared to their non-glycosylated counterparts in U87 MG and U343 MGa glioma cells. Both types of dendrimers were effective in reducing tumor spheroid size, demonstrating their potential as drug delivery agents in glioma treatment [44].

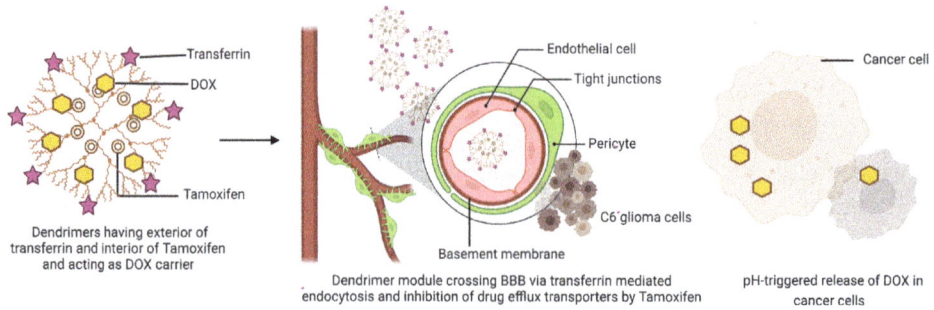

FIGURE 10.2 Targeted dendrimer delivery system for glioma therapy. This illustration depicts the delivery of doxorubicin (DOX) using dendrimers modified with transferrin and tamoxifen for glioma treatment.

10.4.3 NANOSTRUCTURES

Various nanostructures play crucial roles in drug delivery to the brain and tumor sites. Carbon nanotubes (CNTs), with their typical cylindrical structure, can penetrate the BBB and deliver drugs to the brain [45]. Moreover, CNTs exhibit electrical conductivity, enabling them to generate photothermal features [46] and strong Raman signals [47]. These properties make CNTs valuable for tumor-targeted drug delivery systems (DDS) [48], photodynamic therapy (PDT) [48], and photothermal therapy (PTT) [49].

A magnetic DNA-gold carbon nanocage has been developed for telomerase imaging. This nanoprobe enables the monitoring of telomerase activity in tumor cells and induces tumor cell death through a photodynamic–photothermal combinatorial effect [50].

PEGylated oxidized multi-walled carbon nanotubes, functionalized with angiopep-2, possess a high surface area for loading significant amounts of DOX. Angiopep-2 functionalization facilitates the targeting of drugs to brain tumors via low-density lipoprotein receptor-related protein receptors. In mice glioma models, this nanocomposite exhibits anticancer effects and excellent tolerability, demonstrating successful targeting of the ligands in glioma cells [51].

Carbon nanotubes possess remarkable cell-penetrating properties, facilitating the delivery of loaded therapeutic molecules and aiding in photodynamic and photothermal therapy [45]. The human ferritin heavy chain nanocage structure is utilized to carry Paclitaxel, with the drug-loaded nanocage specifically binding to overexpressed transferrin receptor 1 (TfR1) on the BBB and glioma cells. This enhanced binding improves drug accumulation at the target site, resulting in anti-tumor activity in animal glioma models [52].

Another nanocage, composed of apoferritin loaded with DOX, also binds to TfR1 receptors and exhibits significant BBB-crossing and brain tumor-accumulating properties. In mouse models, DOX-loaded apoferritin cages, an iron storage protein, improve overall survival compared to those receiving DOX alone [53].

10.4.4 NANO GELS FOR SUSTAINED RELEASE OF DRUG

Chlorogenic acid (CGA), a nutraceutical found in coffee, exhibits effectiveness in killing glioma cells. However, its rapid clearance and weak stability limit its utility against glioma. To address this challenge, CGA-loaded hydrogels have been developed, which have demonstrated sustained effectiveness [54]. Additionally, a thermoresponsive hydrogel based on polyethylene glycol-calcium phosphate NPs has been engineered to deliver both paclitaxel and temozolomide, inducing autophagy and showing anti-glioma efficacy in C6 tumor-bearing rats [55]. Furthermore, thermoreversible gelation polymer conjugated with DOX has been investigated for pharmacokinetic studies against human glioma cell lines and tumor models in nude mice. Evaluation of its anti-tumor efficacy, using the proliferation marker Ki-67, revealed a reduction in proliferation in the treatment group [56]. In another approach, a gel-temozolomide formulation was injected into the post-resection cavity in the cranium, followed by biodistribution and safety assessments. The formulation was well tolerated and effectively reduced tumor load [57]. These findings collectively suggest that various nanocarriers, characterized by reproducibility, low cost, and extended shelf life, hold promise for efficient drug delivery to brain tumors.

10.5 CONCLUSION

This chapter on brain targeting for drug delivery to combat brain cancer has elucidated several innovative approaches to navigate the complex environment of the BBB and deliver therapeutic agents directly to brain tumors, specifically gliomas. The exploration of various transport mechanisms including nasal, olfactory, and CSF pathways alongside advanced nanotechnologies such as dendrimers, lipid NPs, and nano gels, showcases the breadth of current research and development in this field. We discussed how each transport mechanism and nanotechnology has been tailored to enhance drug bioavailability and efficacy in the brain, overcoming the natural barriers that have traditionally impeded effective brain cancer treatments. The inclusion of receptor-mediated transport, the use of specific targeting ligands, and the design of NPs to exploit natural cellular transport processes highlight the sophistication of current strategies. Moreover, the development of pH-responsive systems, thermos-reversible hydrogels, and other sustained-release formulations are promising advancements that offer controlled release and targeted delivery, which are critical for improving therapeutic outcomes in glioma treatment. These systems are designed to release their therapeutic cargo in response to specific environmental triggers within the tumor microenvironment, which minimizes systemic toxicity and maximizes therapeutic efficacy. The future of brain cancer therapy lies in the continued refinement of these delivery systems and the integration of multidisciplinary research approaches. The ongoing clinical trials and translational research in this domain are critical for bringing these innovative therapies from the laboratory to the bedside, offering hope for improved survival and quality of life for patients with brain cancer.

REFERENCES

[1] Fares, J.; Fares, M.Y.; Khachfe, H.H.; Salhab, H.A.; Fares, Y. Molecular Principles of Metastasis: A Hallmark of Cancer Revisited. *Sig Transduct Target Ther*, 2020, *5*, 1–17.

[2] Bhowmik, A.; Khan, R.; Ghosh, M.K. Blood Brain Barrier: A Challenge for Effectual Therapy of Brain Tumors. *Biomed Res Int*, 2015, *2015*, 320941.

[3] Marei, H.E. Multimodal Targeting of Glioma with Functionalized Nanoparticles. *Cancer Cell Int*, 2022, *22*, 265.

[4] Guerrero-Cázares, H.; Tzeng, S.Y.; Young, N.P.; Abutaleb, A.O.; Quiñones-Hinojosa, A.; Green, J.J. Biodegradable Polymeric Nanoparticles Show High Efficacy and Specificity at DNA Delivery to Human Glioblastoma in vitro and in vivo. *ACS Nano*, 2014, *8*, 5141–5153.

[5] Waite, C.L.; Roth, C.M. PAMAM-RGD Conjugates Enhance SiRNA Delivery through a Multicellular Spheroid Model of Malignant Glioma. *Bioconjug Chem*, 2009, *20*, 1908–1916.

[6] Majumder, P.; Baxa, U.; Walsh, S.T.R.; Schneider, J.P. Design of a Multicompartment Hydrogel That Facilitates Time-Resolved Delivery of Combination Therapy and Synergized Killing of Glioblastoma. *Angew Chem Int Ed Engl*, 2018, *57*, 15040–15044.

[7] Begines, B.; Ortiz, T.; Pérez-Aranda, M.; Martínez, G.; Merinero, M.; Argüelles-Arias, F.; Alcudia, A. Polymeric Nanoparticles for Drug Delivery: Recent Developments and Future Prospects. *Nanomaterials (Basel)*, 2020, *10*, 1403.

[8] Selvaraj, K.; Gowthamarajan, K.; Karri, V.V.S.R. Nose to Brain Transport Pathways an Overview: Potential of Nanostructured Lipid Carriers in Nose to Brain Targeting. *Artif Cells Nanomed Biotechnol*, 2018, *46*, 2088–2095.

[9] Maynard, R.L.; Downes, N. Nasal Cavity. In: *Anatomy and Histology of the Laboratory Rat in Toxicology and Biomedical Research*; Elsevier, 2019; pp. 109–121.

[10] Arora, P.; Sharma, S.; Garg, S. Permeability Issues in Nasal Drug Delivery. *Drug Discov Today*, 2002, *7*, 967–975.

[11] Leal, J.; Smyth, H.D.C.; Ghosh, D. Physicochemical Properties of Mucus and Their Impact on Transmucosal Drug Delivery. *Int J Pharm*, 2017, *532*, 555–572.

[12] Mittal, H.; Jindal, R.; Kaith, B.S.; Maity, A.; Ray, S.S. Synthesis and Flocculation Properties of Gum Ghatti and Poly(Acrylamide-Co-Acrylonitrile) Based Biodegradable Hydrogels. *Carbohydr Polym*, 2014, *114*, 321–329.

[13] Illum, L. Transport of Drugs from the Nasal Cavity to the Central Nervous System. *Eur J Pharm Sci*, 2000, *11*, 1–18.

[14] Upadhaya, P.G.; Pulakkat, S.; Patravale, V.B. Nose-to-Brain Delivery: Exploring Newer Domains for Glioblastoma Multiforme Management. *Drug Deliv Transl Res*, 2020, *10*, 1044–1056.

[15] Sabir, F.; Ismail, R.; Csoka, I. Nose-to-Brain Delivery of Antiglioblastoma Drugs Embedded into Lipid Nanocarrier Systems: Status Quo and Outlook. *Drug Discov Today*, 2020, *25*, 185–194.

[16] Jeong, S.-H.; Jang, J.-H.; Lee, Y.-B. Drug Delivery to the Brain via the Nasal Route of Administration: Exploration of Key Targets and Major Consideration Factors. *J Pharm Investig*, 2023, *53*, 119–152.

[17] Kanazawa, T.; Taki, H.; Okada, H. Nose-to-Brain Drug Delivery System with Ligand/Cell-Penetrating Peptide-Modified Polymeric Nano-Micelles for Intracerebral Gliomas. *Eur J Pharm Biopharm*, 2020, *152*, 85–94.

[18] Kou, D.; Gao, Y.; Li, C.; Zhou, D.; Lu, K.; Wang, N.; Zhang, R.; Yang, Z.; Zhou, Y.; Chen, L.; Ge, J.; Zeng, J.; Gao, M. Intranasal Pathway for Nanoparticles to Enter the Central Nervous System. *Nano Lett*, 2023.

[19] Johnston, M.; Zakharov, A.; Papaiconomou, C.; Salmasi, G.; Armstrong, D. Evidence of Connections between Cerebrospinal Fluid and Nasal Lymphatic Vessels in Humans, Non-Human Primates and Other Mammalian Species. *Cerebrospinal Fluid Res*, 2004, *1*, 2.

[20] Khan, A.R.; Liu, M.; Khan, M.W.; Zhai, G. Progress in Brain Targeting Drug Delivery System by Nasal Route. *J Control Release*, 2017, *268*, 364–389.

[21] Tashima, T. Brain Cancer Chemotherapy Through a Delivery System across the Blood-Brain Barrier into the Brain Based on Receptor-Mediated Transcytosis Using Monoclonal Antibody Conjugates. *Biomedicines*, 2022, *10*, 1597.

[22] Perry, J.; Chambers, A.; Spithoff, K.; Laperriere, N. Gliadel Wafers in the Treatment of Malignant Glioma: A Systematic Review. *Curr Oncol*, 2007, *14*, 189–194.

[23] Fang, F.; Zou, D.; Wang, W.; Yin, Y.; Yin, T.; Hao, S.; Wang, B.; Wang, G.; Wang, Y. Non-Invasive Approaches for Drug Delivery to the Brain Based on the Receptor Mediated Transport. *Mater Sci Eng C Mater Biol Appl*, 2017, *76*, 1316–1327.

[24] Pardridge, W.M. Biologic TNFα-Inhibitors That Cross the Human Blood-Brain Barrier. *Bioeng Bugs*, 2010, *1*, 231–234.

[25] Eavarone, D.A.; Yu, X.; Bellamkonda, R.V. Targeted Drug Delivery to C6 Glioma by Transferrin-Coupled Liposomes. *J Biomed Mater Res*, 2000, *51*, 10–14.

[26] Wang, S.; Reinhard, S.; Li, C.; Qian, M.; Jiang, H.; Du, Y.; Lächelt, U.; Lu, W.; Wagner, E.; Huang, R. Antitumoral Cascade-Targeting Ligand for IL-6 Receptor-Mediated Gene Delivery to Glioma. *Mol Ther*, 2017, *25*, 1556–1566.

[27] Kusaka, S.; Morizane, Y.; Tokumaru, Y.; Tamaki, S.; Maemunah, I.R.; Akiyama, Y.; Sato, F.; Murata, I. Boron Delivery to Brain Cells via Cerebrospinal Fluid (CSF) Circulation for BNCT in a Rat Melanoma Model. *Biology (Basel)*, 2022, *11*, 397.

[28] Maher, F.; Vannucci, S.J.; Simpson, I.A. Glucose Transporter Proteins in Brain. *FASEB J*, 1994, *8*, 1003–1011.

[29] Haugeto, O.; Ullensvang, K.; Levy, L.M.; Chaudhry, F.A.; Honoré, T.; Nielsen, M.; Lehre, K.P.; Danbolt, N.C. Brain Glutamate Transporter Proteins Form Homomultimers. *J Biol Chem*, 1996, *271*, 27715–27722.

[30] Qian, Z.M.; Wang, Q. Expression of Iron Transport Proteins and Excessive Iron Accumulation in the Brain in Neurodegenerative Disorders. *Brain Res Brain Res Rev*, 1998, *27*, 257–267.

[31] Lyubartseva, G.; Smith, J.L.; Markesbery, W.R.; Lovell, M.A. Alterations of Zinc Transporter Proteins ZnT-1, ZnT-4 and ZnT-6 in Preclinical Alzheimer's Disease Brain. *Brain Pathol*, 2010, *20*, 343–350.

[32] Thiebaut, F.; Tsuruo, T.; Hamada, H.; Gottesman, M.M.; Pastan, I.; Willingham, M.C. Immunohistochemical Localization in Normal Tissues of Different Epitopes in the Multidrug Transport Protein P170: Evidence for Localization in Brain Capillaries and Crossreactivity of One Antibody with a Muscle Protein. *J Histochem Cytochem*, 1989, *37*, 159–164.

[33] Haar, C.P.; Hebbar, P.; Wallace, G.C.; Das, A.; Vandergrift, W.A.; Smith, J.A.; Giglio, P.; Patel, S.J.; Ray, S.K.; Banik, N.L. Drug Resistance in Glioblastoma: A Mini Review. *Neurochem Res*, 2012, *37*, 1192–1200.

[34] Bronger, H.; König, J.; Kopplow, K.; Steiner, H.-H.; Ahmadi, R.; Herold-Mende, C.; Keppler, D.; Nies, A.T. ABCC Drug Efflux Pumps and Organic Anion Uptake Transporters in Human Gliomas and the Blood-Tumor Barrier. *Cancer Res*, 2005, *65*, 11419–11428.

[35] Scioli Montoto, S.; Muraca, G.; Ruiz, M.E. Solid Lipid Nanoparticles for Drug Delivery: Pharmacological and Biopharmaceutical Aspects. *Front Mol Biosci*, 2020, *7*, 587997.

[36] Müller, R.H.; Radtke, M.; Wissing, S.A. Solid Lipid Nanoparticles (SLN) and Nanostructured Lipid Carriers (NLC) in Cosmetic and Dermatological Preparations. *Adv Drug Deliv Rev*, 2002, *54*(Suppl 1), S131–155.

[37] Müller, R.H.; Radtke, M.; Wissing, S.A. Nanostructured Lipid Matrices for Improved Microencapsulation of Drugs. *Int J Pharm*, 2002, *242*, 121–128.

[38] Pardridge, W.M. Drug Transport across the Blood-Brain Barrier. *J Cereb Blood Flow Metab*, 2012, *32*, 1959–1972.

[39] Zensi, A.; Begley, D.; Pontikis, C.; Legros, C.; Mihoreanu, L.; Wagner, S.; Büchel, C.; von Briesen, H.; Kreuter, J. Albumin Nanoparticles Targeted with Apo E Enter the CNS by Transcytosis and Are Delivered to Neurones. *J Control Release*, 2009, *137*, 78–86.

[40] Rastogi, V.; Yadav, P.; Porwal, M.; Sur, S.; Verma, A. Dendrimer as Nanocarrier for Drug Delivery and Drug Targeting Therapeutics: A Fundamental to Advanced Systematic Review. *Int J Polym Mater Polym Biomater*, 2022, 1–23.

[41] Du, X.; Shi, B.; Liang, J.; Bi, J.; Dai, S.; Qiao, S.Z. Developing Functionalized Dendrimer-like Silica Nanoparticles with Hierarchical Pores as Advanced Delivery Nanocarriers. *Adv Mater*, 2013, *25*, 5981–5985.

[42] Katare, Y.K.; Daya, R.P.; Sookram Gray, C.; Luckham, R.E.; Bhandari, J.; Chauhan, A.S.; Mishra, R.K. Brain Targeting of a Water Insoluble Antipsychotic Drug Haloperidol via the Intranasal Route Using PAMAM Dendrimer. *Mol Pharm*, 2015, *12*, 3380–3388.

[43] Li, Y.; He, H.; Jia, X.; Lu, W.-L.; Lou, J.; Wei, Y. A Dual-Targeting Nanocarrier Based on Poly(Amidoamine) Dendrimers Conjugated with Transferrin and Tamoxifen for Treating Brain Gliomas. *Biomaterials*, 2012, *33*, 3899–3908.

[44] Dhanikula, R.S.; Argaw, A.; Bouchard, J.-F.; Hildgen, P. Methotrexate Loaded Polyether-Copolyester Dendrimers for the Treatment of Gliomas: Enhanced Efficacy and Intratumoral Transport Capability. *Mol Pharm*, 2008, *5*, 105–116.

[45] Guo, Q.; Shen, X.-T.; Li, Y.-Y.; Xu, S.-Q. Carbon Nanotubes-Based Drug Delivery to Cancer and Brain. *J Huazhong Univ Sci Technolog Med Sci*, 2017, *37*, 635–641.

[46] Chen, D.; Wang, C.; Nie, X.; Li, S.; Li, R.; Guan, M.; Liu, Z.; Chen, C.; Wang, C.; Shu, C.; Wan, L. Photoacoustic Imaging Guided Near-Infrared Photothermal Therapy Using Highly Water-Dispersible Single-Walled Carbon Nanohorns as Theranostic Agents. *Advanced Functional Materials*, 2014, *24*, 6621–6628.

[47] Zhang, J.; Perrin, M.L.; Barba, L.; Overbeck, J.; Jung, S.; Grassy, B.; Agal, A.; Muff, R.; Brönnimann, R.; Haluska, M.; Roman, C.; Hierold, C.; Jaggi, M.; Calame, M. High-Speed Identification of Suspended Carbon Nanotubes Using Raman Spectroscopy and Deep Learning. *Microsyst Nanoeng*, 2022, *8*, 19.

[48] Tang, L.; Xiao, Q.; Mei, Y.; He, S.; Zhang, Z.; Wang, R.; Wang, W. Insights on Functionalized Carbon Nanotubes for Cancer Theranostics. *J Nanobiotechnology*, 2021, *19*, 423.

[49] Zhao, Y.; Zhao, T.; Cao, Y.; Sun, J.; Zhou, Q.; Chen, H.; Guo, S.; Wang, Y.; Zhen, Y.; Liang, X.-J.; Zhang, S. Temperature-Sensitive Lipid-Coated Carbon Nanotubes for Synergistic Photothermal Therapy and Gene Therapy. *ACS Nano*, 2021, *15*, 6517–6529.

[50] Shen, F.; Zhang, C.; Cai, Z.; Wang, J.; Zhang, X.; Machuki, J.O.; Cui, L.; Li, S.; Gao, F. Carbon Nanocage/Fe$_3$O$_4$/DNA-Based Magnetically Targeted Intracellular Imaging of Telomerase via Catalyzed Hairpin Assembly and Photodynamic-Photothermal Combination Therapy of Tumor Cells. *ACS Appl Mater Interfaces*, 2020, *12*, 53624–53633.

[51] Ren, J.; Shen, S.; Wang, D.; Xi, Z.; Guo, L.; Pang, Z.; Qian, Y.; Sun, X.; Jiang, X. The Targeted Delivery of Anticancer Drugs to Brain Glioma by PEGylated Oxidized Multi-Walled Carbon Nanotubes Modified with Angiopep-2. *Biomaterials*, 2012, *33*, 3324–3333.

[52] Liu, W.; Lin, Q.; Fu, Y.; Huang, S.; Guo, C.; Li, L.; Wang, L.; Zhang, Z.; Zhang, L. Target Delivering Paclitaxel by Ferritin Heavy Chain Nanocages for Glioma Treatment. *J Control Release*, 2020, *323*, 191–202.

[53] Chen, Z.; Zhai, M.; Xie, X.; Zhang, Y.; Ma, S.; Li, Z.; Yu, F.; Zhao, B.; Zhang, M.; Yang, Y.; Mei, X. Apoferritin Nanocage for Brain Targeted Doxorubicin Delivery. *Mol Pharm*, 2017, *14*, 3087–3097.

[54] Zhou, H.; Chen, D.; Gong, T.; He, Q.; Guo, C.; Zhang, P.; Song, X.; Ruan, J.; Gong, T. Chlorogenic Acid Sustained-Release Gel for Treatment of Glioma and Hepatocellular Carcinoma. *Eur J Pharm Biopharm*, 2021, *166*, 103–110.

[55] Ding, L.; Wang, Q.; Shen, M.; Sun, Y.; Zhang, X.; Huang, C.; Chen, J.; Li, R.; Duan, Y. Thermoresponsive Nanocomposite Gel for Local Drug Delivery to Suppress the Growth of Glioma by Inducing Autophagy. *Autophagy*, 2017, *13*, 1176–1190.

[56] Arai, T.; Joki, T.; Akiyama, M.; Agawa, M.; Mori, Y.; Yoshioka, H.; Abe, T. Novel Drug Delivery System Using Thermoreversible Gelation Polymer for Malignant Glioma. *J Neurooncol*, 2006, *77*, 9–15.

[57] Akbar, U.; Jones, T.; Winestone, J.; Michael, M.; Shukla, A.; Sun, Y.; Duntsch, C. Delivery of Temozolomide to the Tumor Bed via Biodegradable Gel Matrices in a Novel Model of Intracranial Glioma with Resection. *J Neurooncol*, 2009, *94*, 203–212.

11 Phytochemicals for Brain Tumor Therapy
Challenges and Innovations

Zirong Fan[1*], Rekha Khandia[2], and Pankaj Gurjar[3]
[1]Department of Neurosurgery, Shengjing Hospital of China Medical University, Shenyang, Liaoning, China
[2]Department of Biochemistry and Genetics, Barkatullah University, Bhopal MP, India
[3]Department of Science and Engineering, Novel Global Community Educational Foundation, Hebersham, Australia
*Corresponding author

11.1 INTRODUCTION

Brain tumors, though rare, often result in significant mortality and morbidity, leading to a poor quality of life [1]. Among these tumors, glioblastoma multiforme (GBM) stands out as the most aggressive form originating from glial cells [2]. Standard treatment for GBM typically involves therapeutic intervention and surgical resection, followed by temozolomide (TMZ) therapy. TMZ treatment, in combination with radiotherapy, has been shown to extend the average lifespan of patients. Specifically, the combination therapy increases progression-free survival from 3.9 months to 5.3 months and overall survival from 7.7 months (with radiotherapy alone) to 13.5 months [3]. Additionally, Bevacizumab, a humanized anti-vascular endothelial growth factor monoclonal antibody, is indicated for angiogenesis inhibition [4]. It has been approved by the US Food and Drug Administration for the treatment of colorectal, lung, breast, brain (GBM), and renal cell cancers, resulting in improved survival rates. However, it is associated with severe side effects such as intestinal perforation, hemorrhage, and delayed wound healing [5].

The roots of *Bupleurum chinense* and *Bupleurum scorzonerifolium* wild plants contain a toxic compound, saikosaponin D, which exhibits anticancer activity. It downregulates the PI3K/Akt and ERK pathways while upregulating c-Jun N-terminal kinases (JNK), leading to enhanced apoptosis in U87 glioma cells [6]. In C6 cells, it has been shown to induce differentiation and inhibit growth [7]. Additionally, acetone extracts from these plants induce cell cycle arrest in the G2/M phase and stimulate tubulin polymerization in A549 cells [8]. Furthermore, ethanol extracts from *Securidacalonge pedunculata, Andirainermis subsp. rooseveltii, Annonasenegalensis, Carissa edulis*, and *Parinari polyandra* have been evaluated for their activity against brain tumor cells in vitro. All of these extracts demonstrate a significant reduction in brain tumor cells, with *S. longepedunculata* being the most effective at the lowest concentration, particularly in inhibiting the invasion of the U251 cell line [9]. This chapter discusses various plant-derived materials for cancer therapy, with Figure 11.1 illustrating plant extract-based cancer therapy.

11.2 PLANT SAP-DERIVED EXTRACELLULAR VESICLES

Extracellular vesicles (EVs) play crucial roles in cell-to-cell communication, akin to mammalian exosomes, and have the potential to efficiently transfer plant metabolites into target cells for thera peutic effects [10, 11]. These EVs are nano-sized and contain various therapeutic molecules [12].

DOI: 10.1201/9781003519706-11

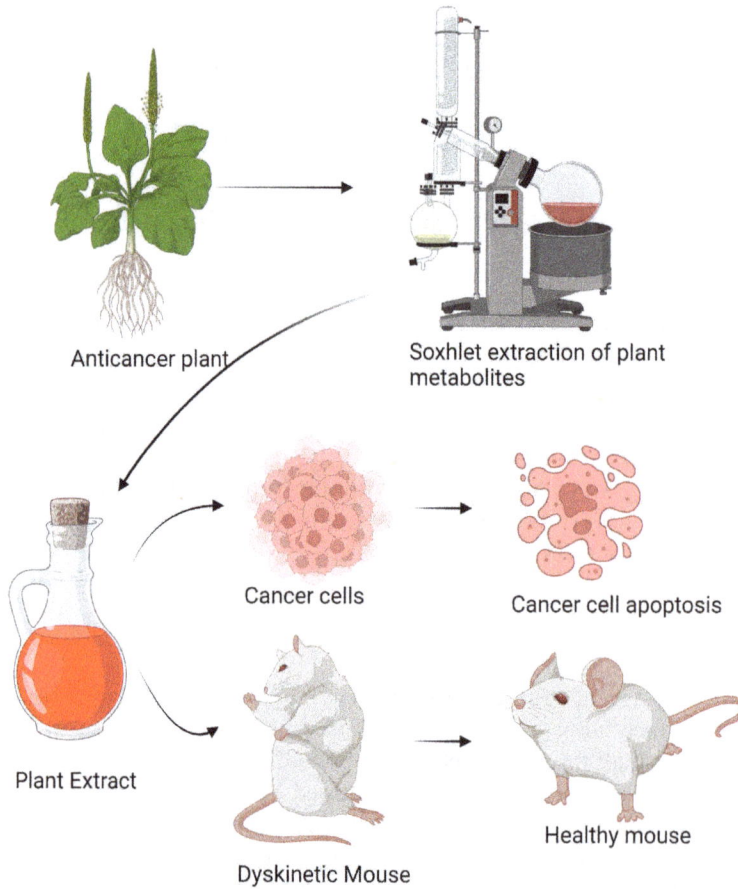

FIGURE 11.1 Extraction and testing of anticancer compounds from medicinal plants. This figure illustrates the process of deriving anticancer compounds from a medicinal plant through various stages.

Plant-derived EVs are rich in miRNA, non-coding small RNAs, and metabolites related to growth, differentiation, and metabolism [13]. They are also integral to plant defense mechanisms, aiding in the prevention of pathogen attacks [14].

Through the use of differential ultracentrifugation, nanovesicles have been isolated from lemon juice (*Citrus limon* L.), which have been found to inhibit tumor cell growth. Lemon nanovesicles induce TRAIL-mediated apoptotic cell death, suggesting the potential for edible plant-derived therapeutic strategies [15]. Additionally, lipid nanovesicles encapsulating plant extracts, known as ethosomes or phytosomes, have demonstrated efficacy in wound healing [16].

Edible plant-derived exosome-like nanoparticles (EPDENs) have been isolated and characterized for their biological activities, comprising proteins, lipids, and microRNAs. In an experiment involving ginger EPDENs transfected into macrophages, expression of the antioxidant gene and inflammatory cytokines was observed. This experiment serves as a proof of concept, demonstrating interspecies communication through EPDENs [17].

EVs derived from *Dendropanax morbifera* and *Pinus densiflora* saps have demonstrated efficacy in tumor treatment [18]. Additionally, a grapefruit-derived nano vector loaded with miR17, functionalized with folic acid and coated with polyethylenimine, has been developed. This modification enhances the RNA carrying capacity while reducing polyethylenimine toxicity. When administered intranasally to mice brains, these miR17-carrying nanovesicles were engulfed by GL-26 tumor cells, resulting in reduced tumor growth [19].

11.3   RESVERATROL

Resveratrol (RES) is a polyphenol found in various plants such as grapes, wine, peanuts, soy, and Itadori tea [20]. Plants produce RES in response to stressors like injury, UV radiation, or fungal infection, utilizing the phenylalanine pathway for biosynthesis [21]. TMZ is a common drug used to treat GBM; however, glioma cells often develop resistance to it [22]. Similarly, resistance to RES can also occur, but when combined, they significantly affect the expression of O6-methylguanine-DNA methyltransferase (MGMT) (Figure 11.2) [3].

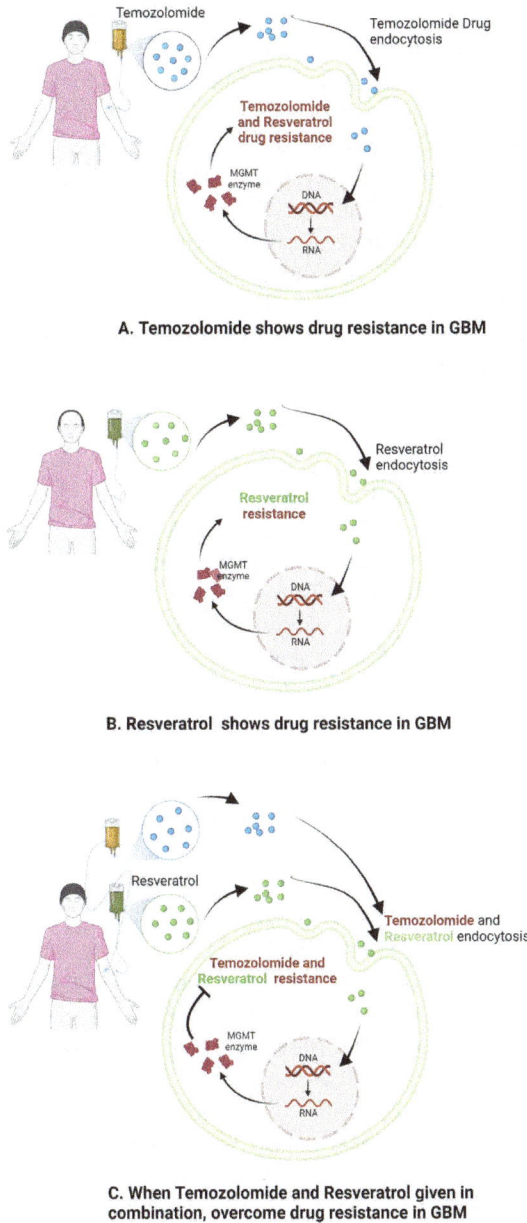

FIGURE 11.2   Overcoming drug resistance in glioblastoma multiforme (GBM) with temozolomide and resveratrol combination therapy. This figure illustrates the mechanisms of drug resistance and the potential overcoming of such resistance in GBM through combination therapy.

Surgical resection or radiotherapy alone is often used for GBM treatment, but it remains pallia-tive due to the radioresistant nature of GBM stem cells. The effects of RES combined with radio-therapy were evaluated in radioresistant GBM cell lines. This combination treatment downregulated the expression of the neural stem cell marker CD133 and enhanced the efficacy of radiosensitization by attenuating radiation-induced DNA damage repair. Therefore, RES may be utilized to augment the effects of radiotherapy against GBM stem cells [23]. The effects of RES on the proliferation and viability of seven patient-derived GBM stem cell lines were also assessed.

RES has shown effectiveness in inhibiting tumor cell proliferation and motility; however, its impact varies concerning c-Myc and β-catenin expression [24]. With the ability to penetrate the blood–brain barrier (BBB), it exhibits antioxidant, anti-inflammatory, anti-diabetic, and anti-tumor effects, attributed to its influence on various signaling pathways [25].

The over-activation of PI3K/AKT pathways contributes to high GBM cell survival, while increased expression of the p-glycoprotein transporter is linked to a poor prognosis. RES's activity in inhibiting PI3K pathways may thus hold therapeutic promise for GBMs. Studies on human DOX-sensitive and DOX-resistant U87MG cells have demonstrated RES's ability to overcome drug resis-tance, as evidenced by Rhodamines 123 uptake. Moreover, it inhibits PI3K signaling and activates P-glycoprotein, rendering GBM cells more sensitive to chemotherapeutic agents [26].

When administered alongside TMZ, RES induces cell cycle arrest and astrocyte differentiation, as indicated by the overexpression of the differentiation marker glial fibrillary acid protein [27]. This combination treatment increases mTOR signaling, reduces the anti-apoptotic protein Bcl-2, and significantly decreases tumor volume [28].

11.4 QUERCETIN

Quercetin, a flavonol abundantly found in plants, possesses potent antioxidant, anti-inflammatory, antiviral, and anticancer properties with low toxicity, making it an attractive candidate for thera-peutic applications [29]. Its antiproliferative effects on cancer cells involve multiple mechanisms, including the regulation of the PI3K/Akt/mTOR pathway, IL6/STAT pathway, heat shock protein expression, and modulation of intracellular pH [30]. Quercetin inhibits mesenchymal transition via the GSK-3β/β-catenin/ZEB1 signaling pathway and effectively acts as an anti-GBM drug [31].

Treatment of GBM cells with quercetin for 24 hours at concentrations below the IC50 results in the inhibition of cell proliferation, invasion, and angiogenesis, possibly through the downregula-tion of VEGFA, MMP9, and MMP2 proteins [32]. Studies on U87MG and T98G cell lines and GBM mouse models have demonstrated the anticancer effects of quercetin, including inhibition of proliferation, induction of apoptosis, and reduction of glycolytic metabolism [33]. The Warburg effect, wherein cancer cells rely on glycolysis despite sufficient oxygen availability, promotes tumor growth [34]. Quercetin's inhibition of glycolytic metabolism suggests its potential for treating a wide range of cancers. Moreover, quercetin inhibits the IL-6-induced STAT3 pathway in T98G and U87 GBM cells, leading to reduced proliferation and migration of GBM cells, along with modu-lation of cyclin D1 and matrix metalloproteinase-2 (MMP-2) expression [35]. A nano hydrogel combining quercetin and TMZ shows enhanced internalization and cytotoxicity in GBM cells. This nanocomposite targets the CD44 receptor, internalizes via caveolae-dependent mechanisms, and modulates the tumor microenvironment to reduce tumor cell proliferation [36]. Co-administration of quercetin and TMZ induces apoptosis in T98G and MOGGCCM cells through altered nuclei shape and expression of Hsp27 and Hsp72, along with activation of caspase-9 and caspase-3 [37]. Furthermore, quercetin combined with sodium butyrate increases apoptosis in GBM cell lines by reducing the expression of PARP-1 and survivin, thus presenting a promising chemotherapeutic agent when used in conjunction with conventional therapies [38].

Five malignant glioma cell lines, including U87-MG, U251, A172, LN229, and U373, were sub-jected to treatment to induce tumor cell death. Following treatment, a significant decrease in sur-vivin levels was observed in most of the GBM cell lines, primarily through proteasomal degradation.

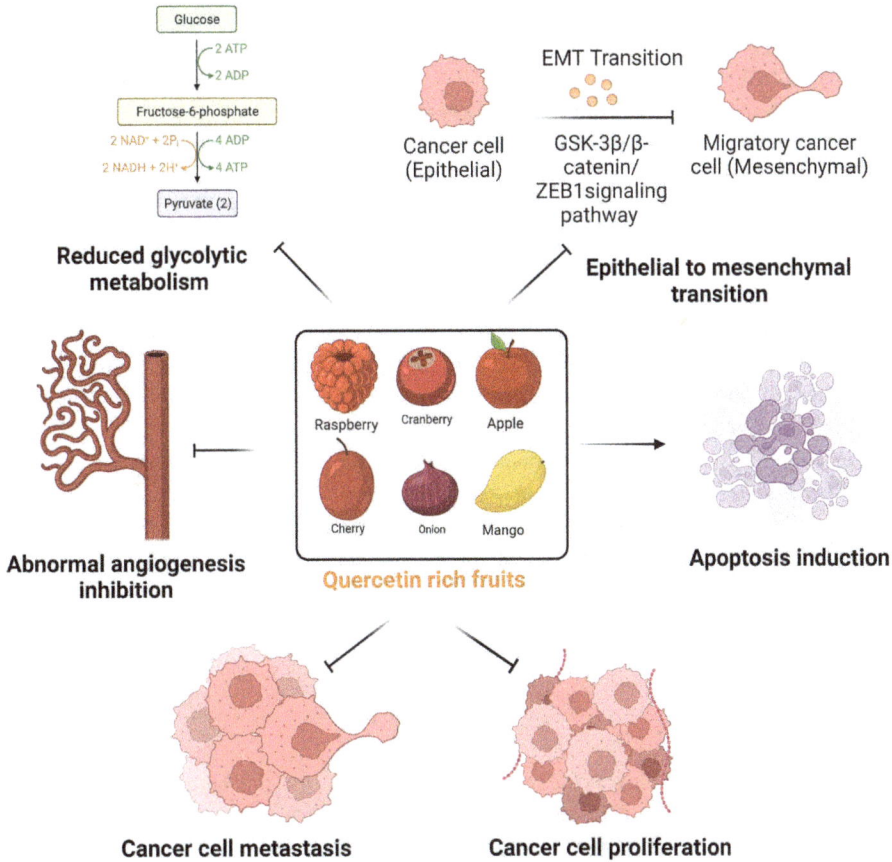

FIGURE 11.3 Impact of quercetin-rich fruits on cancer cell dynamics. This figure provides a comprehensive overview of the biochemical and physiological impacts of quercetin-rich fruits on cancer cell processes.

Quercetin has demonstrated efficacy in sensitizing glioma cells to death-receptor-mediated apoptosis, offering potential therapeutic benefits [39]. Chloroquine, an antimalarial drug, when combined with quercetin, induces the expansion of autolysosomes and lysosomes, leading to cellular death (Figure 11.3). Additionally, mitochondrial uniporter (MCU)-mediated Ca^{2+} influx into mitochondria further enhances the cytotoxic effects on GBM cells [40]. Despite its promise in brain tumor treatment, quercetin faces challenges related to poor oral bioavailability (<2%) and limited brain permeability. To address these limitations, quercetin-loaded nano lipidic carriers (NLCs) have been developed, incorporating phospholipids and tocopherol acetate to improve delivery and BBB penetration. Encapsulation enhances uptake, residence time, and reduces drug clearance, making nanocapsulated quercetin a promising platform for brain delivery [41].

11.4.1 CAMPTOTHECIN

Camptothecin (CPT), an alkaloid derived from the *Camptotheca acuminata* tree, functions by forming a complex with topoisomerase I and DNA, inducing DNA damage and apoptosis. While highly proliferative tumors are sensitive to CPT, its effectiveness is hindered by water insolubility and rapid hydrolysis, limiting its permeability through the BBB [42]. To address these issues, CPT derivatives have been synthesized, with irinotecan (CPT 11) and topotecan being the only two approved for clinical use due to their ability to cross the BBB.

For targeted delivery and enhanced therapeutic effects, CPT has been conjugated to 5-aminolevulinate, exhibiting promising photodynamic properties and reduced side effects. This approach has shown both in vitro and in vivo anti-tumor activities, suggesting its potential in GBM therapy [43]. Additionally, nasal administration of CPT with cell-penetrating peptide and copolymers-based micelles demonstrated high toxicity in rat GBM cells and increased survival rates, highlighting the efficacy of this delivery method [44].

Furthermore, poly(lactic-co-glycolic acid) microspheres containing CPT have been utilized for retention and sustained release in a rat model of malignant glioma. This approach prolonged survival compared to traditional buffer solution administration, indicating the potential of microspheres for sustained drug delivery in brain tumor therapy [45].

Furthermore, the cytotoxicity of the CPT molecule may be enhanced by combining it with staurosporine, which synergistically increases cytotoxicity and apoptosis in U251 and DAOY cells. Such therapies demonstrate the novel use of cell cycle checkpoints to inhibit cancer progression using natural plant-derived molecules [46].

To enhance the bioavailability of CPT, nanoparticles can also be utilized. However, CPT is active at pH levels below 5 (in its lactone form) and becomes irreversibly inactivated at basic pH. To overcome this challenge, cyclodextrin-based nanoparticle formulations have been developed, demonstrating sustained release for up to 12 days and slightly improving mean survival time in animal models [47]. In another study by Lin et al. (2016), CPT was conjugated with cyclodextrin-polyethylene glycol and investigated for its ability to cross tight-junction barriers in the brain. This conjugate showed promise in preventing hypoxia and angiogenesis in a glioma mouse tumor model [48].

11.5 ANTHOCYANINS

Anthocyanins, water-soluble flavonoid pigments responsible for the blue, purple, red, and intermediate colors of various plant parts, have garnered attention for their antioxidant properties [49]. Canadian elderberry (*Sambucus nigra*) is particularly rich in these antioxidant anthocyanins. In a study, two Canadian elderberry cultivar extracts were evaluated against human brain tumor cells and brain microvascular endothelial cells. The extracts showed inhibition of cell cycle progression through modulation of cell cycle checkpoints and apoptosis. Additionally, cranberry extract demonstrated efficacy in targeting both brain tumor cells and microvasculature [50].

Anthocyanins such as aglycons, cyanidin, delphinidin, and petunidin have been identified as potent inhibitors of GBM cell migration [51]. Epithelial-to-mesenchymal transition (EMT) is a reversible process in which a cell's morphology, adhesion, and migration pattern are altered, often induced by factors like Transforming Growth Factor-beta (TGF-β). Studies have assessed the impact of anthocyanidins like cyanidin, delphinidin, malvidin, pelargonidin, and petunidin on EMT markers such as fibronectin, snail, Smad2, ERK, and JNK. Anthocyanidins downregulated these markers, inhibiting U-87 MG cell migration, suggesting a potential role of anthocyanins in dietary-based nutraceuticals against brain tumors [52]. Furthermore, the effects of anthocyanidins on TGF-β-induced EMT vary depending on treatment conditions, with delphinidin showing potent inhibition of TGF-β Smad and non-Smad signaling pathways, displaying tumor preventive roles [53]. Drug resistance is a significant challenge in brain tumor treatment. Cyanidin-3-O-glucoside (C3G), a derivative of cyanidin, was evaluated in TMZ-resistant cells. C3G upregulated miR-214-5p, helping overcome drug resistance and enhancing the cytotoxicity of TMZ, leading to significant inhibition in LN-18/TR tumor growth [54].

Dracorhodin perchlorate, a red pigment, has been observed to induce apoptosis in cancer cells, as evidenced by MTT assay results in glioma U87MG and T98G cells. It functions by blocking the G1/S phase transition, which is associated with the upregulation of p53 and p21 protein expression. Additionally, it disrupts the cellular mitochondrial membrane potential, leading to the liberation of cytochrome c and increased expression levels of Bim and Bax apoptotic proteins. These findings suggest that Dracorhodin perchlorate pigment plays a role in cell cycle arrest and the induction of apoptosis in brain tumor cells.

11.6 CUCURBITACINS

Cucurbitacins, tetracyclic triterpenoids found in plants of the Cucurbitaceae family, possess various pharmacological properties [55]. Cucurbitacin E (CuE), extracted from the stems of Cucumis melo L., induces cell cycle arrest at the G2/M phase in GBM 8401 cells [56]. Genome profiling studies using microarray analysis have revealed that CuE treatment leads to the upregulation of 558 genes and the downregulation of 1,354 genes. This analysis suggests that CuE has the potential to inhibit the proliferation of glioma cells, making it a promising candidate for glioma therapy. Similar anti-tumor effects of CuE have been observed in GBM 8401 and U-87-MG cancer cells, where its activity is associated with a cell cycle arrest in the G2/M phase [56].

Cucurbitacin B, derived from the leaves of *Tunisian Ecballium elaterium*, demonstrates anti-integrin activity against α5β1 integrin and inhibits angiogenesis, a critical process for cancer progression [57]. Remarkably, it exhibits tolerance even at higher concentrations, suggesting a novel tumor-targeting mechanism as an anti-integrin drug.

Cucurbitacin I serves as a selective inhibitor of JAK2/STAT3 and induces autophagy and apoptosis in GBM cells [58]. Furthermore, incubation with Cucurbitacin I leads to the upregulation of Beclin 1 in GBM cells, providing evidence for its efficacy in apoptosis and autophagy induction. This proof of concept is supported by studies conducted in xenograft models, positioning Cucurbitacin I as a promising candidate for plant-based therapies in brain cancer patients. Additionally, treatment with Cucurbitacin I results in G2/M accumulation, DNA endoreduplication, and the formation of multipolar mitotic spindles in malignant gliomas. Moreover, it downregulates Aurora kinase A, Aurora kinase B, and survivin, further highlighting its potential therapeutic benefits in combating brain cancer [59].

11.7 CONCLUSION

This chapter explored the potential of phytochemicals in treating brain tumors, especially GBM. It discussed various plant-derived compounds like RES, quercetin, and CPT, emphasizing their ability to enhance conventional treatments such as TMZ and radiotherapy through mechanisms like overcoming drug resistance and inducing apoptosis. The emerging field of plant-derived EVs offers a promising delivery method that could improve the efficacy and targeting of these phytochemicals. While further research and clinical trials are needed to fully validate their therapeutic benefits and safety, the evidence suggests that phytochemicals could significantly improve treatment outcomes for brain tumor patients. Continued investigation into these natural compounds and their integration into clinical practice is crucial for developing more effective and precise therapies for brain tumors, potentially leading to better patient outcomes.

REFERENCES

[1] McNeill, K.A. Epidemiology of Brain Tumors. *Neurol Clin*, 2016, *34*, 981–998.

[2] Shi, F.; Guo, H.; Zhang, R.; Liu, H.; Wu, L.; Wu, Q.; Liu, J.; Liu, T.; Zhang, Q. The PI3K Inhibitor GDC-0941 Enhances Radiosensitization and Reduces Chemoresistance to Temozolomide in GBM Cell Lines. *Neuroscience*, 2017, *346*, 298–308.

[3] Perry, J.R.; Laperriere, N.; O'Callaghan, C.J.; Brandes, A.A.; Menten, J.; Phillips, C.; Fay, M.; Nishikawa, R.; Cairncross, J.G.; Roa, W.; Osoba, D.; Rossiter, J.P.; Sahgal, A.; Hirte, H.; Laigle-Donadey, F.; Franceschi, E.; Chinot, O.; Golfinopoulos, V.; Fariselli, L.; Wick, A.; Feuvret, L.; Back, M.; Tills, M.; Winch, C.; Baumert, B.G.; Wick, W.; Ding, K.; Mason, W.P.; Trial Investigators. Short-Course Radiation Plus Temozolomide in Elderly Patients with Glioblastoma. *N Engl J Med*, 2017, *376*, 1027–1037.

[4] Abdi, F.; Arkan, E.; Eidizadeh, M.; Valipour, E.; Naseriyeh, T.; Gamizgy, Y.H.; Mansouri, K. The Possibility of Angiogenesis Inhibition in Cutaneous Melanoma by Bevacizumab-Loaded Lipid-Chitosan Nanoparticles. *Drug Deliv Transl Res*, 2023, *13*, 568–579.

[5] Park, M.N.; Song, H.S.; Kim, M.; Lee, M.-J.; Cho, W.; Lee, H.-J.; Hwang, C.-H.; Kim, S.; Hwang, Y.; Kang, B.; Kim, B. Review of Natural Product-Derived Compounds as Potent Antiglioblastoma Drugs. *Biomed Res Int*, 2017, *2017*, 8139848.

[6] Li, Y.; Cai, T.; Zhang, W.; Zhu, W.; Lv, S. Effects of Saikosaponin D on Apoptosis in Human U87 Glioblastoma Cells. *Mol Med Rep*, 2017, *16*, 1459–1464.

[7] Tsai, Y.-J.; Chen, I.-L.; Horng, L.-Y.; Wu, R.-T. Induction of Differentiation in Rat C6 Glioma Cells with Saikosaponins. *Phytother Res*, 2002, *16*, 117–121.

[8] Cheng, Y.-L.; Lee, S.-C.; Lin, S.-Z.; Chang, W.-L.; Chen, Y.-L.; Tsai, N.-M.; Liu, Y.-C.; Tzao, C.; Yu, D.-S.; Harn, H.-J. Anti-Proliferative Activity of Bupleurum Scrozonerifolium in A549 Human Lung Cancer Cells in Vitro and in Vivo. *Cancer Lett*, 2005, *222*, 183–193.

[9] Ngulde, S.I.; Sandabe, U.K.; Abounader, R.; Zhang, Y.; Hussaini, I.M. Activities of Some Medicinal Plants on the Proliferation and Invasion of Brain Tumor Cell Lines. *Adv Pharmacol Pharm Sci*, 2020, *2020*, 3626879.

[10] Zhang, M.; Viennois, E.; Xu, C.; Merlin, D. Plant Derived Edible Nanoparticles as a New Therapeutic Approach against Diseases. *Tissue Barriers*, 2016, *4*, e1134415.

[11] Regente, M.; Corti-Monzón, G.; Maldonado, A.M.; Pinedo, M.; Jorrín, J.; de la Canal, L. Vesicular Fractions of Sunflower Apoplastic Fluids Are Associated with Potential Exosome Marker Proteins. *FEBS Lett*, 2009, *583*, 3363–3366.

[12] Karamanidou, T.; Tsouknidas, A. Plant-Derived Extracellular Vesicles as Therapeutic Nanocarriers. *Int J Mol Sci*, 2021, *23*, 191.

[13] Xiao, J.; Feng, S.; Wang, X.; Long, K.; Luo, Y.; Wang, Y.; Ma, J.; Tang, Q.; Jin, L.; Li, X.; Li, M. Identification of Exosome-like Nanoparticle-Derived MicroRNAs from 11 Edible Fruits and Vegetables. *PeerJ*, 2018, *6*, e5186.

[14] Boevink, P.C. Exchanging Missives and Missiles: The Roles of Extracellular Vesicles in Plant-Pathogen Interactions. *J Exp Bot*, 2017, *68*, 5411–5414.

[15] Raimondo, S.; Naselli, F.; Fontana, S.; Monteleone, F.; Lo Dico, A.; Saieva, L.; Zito, G.; Flugy, A.; Manno, M.; Di Bella, M.A.; De Leo, G.; Alessandro, R. Citrus Limon-Derived Nanovesicles Inhibit Cancer Cell Proliferation and Suppress CML Xenograft Growth by Inducing TRAIL-Mediated Cell Death. *Oncotarget*, 2015, *6*, 19514–19527.

[16] Kumar, S.; Kumar, A.; Kumar, N.; Singh, P.; Singh, T.U.; Singh, B.R.; Gupta, P.K.; Thakur, V.K. In Vivo Therapeutic Efficacy of Curcuma Longa Extract Loaded Ethosomes on Wound Healing. *Vet Res Commun*, 2022, *46*, 1033–1049.

[17] Mu, J.; Zhuang, X.; Wang, Q.; Jiang, H.; Deng, Z.-B.; Wang, B.; Zhang, L.; Kakar, S.; Jun, Y.; Miller, D.; Zhang, H.-G. Interspecies Communication between Plant and Mouse Gut Host Cells Through Edible Plant Derived Exosome-Like Nanoparticles. *Mol Nutr Food Res*, 2014, *58*, 1561–1573.

[18] Kim, K.; Yoo, H.J.; Jung, J.-H.; Lee, R.; Hyun, J.-K.; Park, J.-H.; Na, D.; Yeon, J.H. Cytotoxic Effects of Plant Sap-Derived Extracellular Vesicles on Various Tumor Cell Types. *J Funct Biomater*, 2020, *11*, 22.

[19] Zhuang, X.; Teng, Y.; Samykutty, A.; Mu, J.; Deng, Z.; Zhang, L.; Cao, P.; Rong, Y.; Yan, J.; Miller, D.; Zhang, H.-G. Grapefruit-Derived Nanovectors Delivering Therapeutic MiR17 Through an Intranasal Route Inhibit Brain Tumor Progression. *Mol Ther*, 2016, *24*, 96–105.

[20] Burns, J.; Yokota, T.; Ashihara, H.; Lean, M.E.J.; Crozier, A. Plant Foods and Herbal Sources of Resveratrol. *J Agric Food Chem*, 2002, *50*, 3337–3340.

[21] Hasan, M.; Bae, H. An Overview of Stress-Induced Resveratrol Synthesis in Grapes: Perspectives for Resveratrol-Enriched Grape Products. *Molecules*, 2017, *22*, 294.

[22] Singh, N.; Miner, A.; Hennis, L.; Mittal, S. Mechanisms of Temozolomide Resistance in Glioblastoma—A Comprehensive Review. *Cancer Drug Resist*, 2021, *4*, 17–43.

[23] Wang, L.; Long, L.; Wang, W.; Liang, Z. Resveratrol, a Potential Radiation Sensitizer for Glioma Stem Cells Both in Vitro and in Vivo. *J Pharmacol Sci*, 2015, *129*, 216–225.

[24] Cilibrasi, C.; Riva, G.; Romano, G.; Cadamuro, M.; Bazzoni, R.; Butta, V.; Paoletta, L.; Dalprà, L.; Strazzabosco, M.; Lavitrano, M.; Giovannoni, R.; Bentivegna, A. Resveratrol Impairs Glioma Stem Cells Proliferation and Motility by Modulating the Wnt Signaling Pathway. *PLoS One*, 2017, *12*, e0169854.

[25] Ashrafizadeh, M.; Mohammadinejad, R.; Farkhondeh, T.; Samarghandian, S. Protective Effect of Resveratrol against Glioblastoma: A Review. *Anticancer Agents Med Chem*, 2021, *21*, 1216–1227.

[26] Zhang, Y.; Zhang, Z.; Mousavi, M.; Moliani, A.; Bahman, Y.; Bagheri, H. Resveratrol Inhibits Glioblastoma Cells and Chemoresistance Progression Through Blockade P-Glycoprotein and Targeting AKT/PTEN Signaling Pathway. *Chem Biol Interact*, 2023, *376*, 110409.

[27] Liu, Y.; Song, X.; Wu, M.; Wu, J.; Liu, J. Synergistic Effects of Resveratrol and Temozolomide Against Glioblastoma Cells: Underlying Mechanism and Therapeutic Implications. *Cancer Manag Res*, 2020, *12*, 8341–8354.

[28] Yuan, Y.; Xue, X.; Guo, R.-B.; Sun, X.-L.; Hu, G. Resveratrol Enhances the Antitumor Effects of Temozolomide in Glioblastoma via ROS-Dependent AMPK-TSC-MTOR Signaling Pathway. *CNS Neurosci Therapeut*, 2012, *18*, 536–546.

[29] Rauf, A.; Imran, M.; Khan, I.A.; Ur-Rehman, M.; Gilani, S.A.; Mehmood, Z.; Mubarak, M.S. Anticancer Potential of Quercetin: A Comprehensive Review. *Phytother Res*, 2018, *32*, 2109–2130.

[30] Tavana, E.; Mollazadeh, H.; Mohtashami, E.; Modaresi, S.M.S.; Hosseini, A.; Sabri, H.; Soltani, A.; Javid, H.; Afshari, A.R.; Sahebkar, A. Quercetin: A Promising Phytochemical for the Treatment of Glioblastoma Multiforme. *Biofactors*, 2020, *46*, 356–366.

[31] Chen, B.; Li, X.; Wu, L.; Zhou, D.; Song, Y.; Zhang, L.; Wu, Q.; He, Q.; Wang, G.; Liu, X.; Hu, H.; Zhou, W. Quercetin Suppresses Human Glioblastoma Migration and Invasion via GSK3β/β-Catenin/ZEB1 Signaling Pathway. *Front Pharmacol*, 2022, *13*, 963614.

[32] Liu, Y.; Tang, Z.-G.; Yang, J.-Q.; Zhou, Y.; Meng, L.-H.; Wang, H.; Li, C.-L. Low Concentration of Quercetin Antagonizes the Invasion and Angiogenesis of Human Glioblastoma U251 Cells. *Onco Targets Ther*, 2017, *10*, 4023–4028.

[33] Wang, L.; Ji, S.; Liu, Z.; Zhao, J. Quercetin Inhibits Glioblastoma Growth and Prolongs Survival Rate through Inhibiting Glycolytic Metabolism. *Chemotherapy*, 2022, *67*, 132–141.

[34] Yu, L.; Chen, X.; Sun, X.; Wang, L.; Chen, S. The Glycolytic Switch in Tumors: How Many Players Are Involved? *J Cancer*, 2017, *8*, 3430–3440.

[35] Michaud-Levesque, J.; Bousquet-Gagnon, N.; Béliveau, R. Quercetin Abrogates IL-6/STAT3 Signaling and Inhibits Glioblastoma Cell Line Growth and Migration. *Exp Cell Res*, 2012, *318*, 925–935.

[36] Barbarisi, M.; Iaffaioli, R.V.; Armenia, E.; Schiavo, L.; De Sena, G.; Tafuto, S.; Barbarisi, A.; Quagliariello, V. Novel Nanohydrogel of Hyaluronic Acid Loaded with Quercetin Alone and in Combination with Temozolomide as New Therapeutic Tool, CD44 Targeted Based, of Glioblastoma Multiforme. *J Cell Physiol*, 2018, *233*, 6550–6564.

[37] Jakubowicz-Gil, J.; Langner, E.; Bądziul, D.; Wertel, I.; Rzeski, W. Silencing of Hsp27 and Hsp72 in Glioma Cells as a Tool for Programmed Cell Death Induction upon Temozolomide and Quercetin Treatment. *Toxicol Appl Pharmacol*, 2013, *273*, 580–589.

[38] Taylor, M.A. *Synergism of Quercetin and Sodium Butyrate for Controlling Growth of Glioblastoma* (Master's thesis, University of South Carolina). 2017.

[39] Siegelin, M.D.; Reuss, D.E.; Habel, A.; Rami, A.; von Deimling, A. Quercetin Promotes Degradation of Survivin and Thereby Enhances Death-Receptor-Mediated Apoptosis in Glioma Cells. *Neuro Oncol*, 2009, *11*, 122–131.

[40] Jang, E.; Kim, I.Y.; Kim, H.; Lee, D.M.; Seo, D.Y.; Lee, J.A.; Choi, K.S.; Kim, E. Quercetin and Chloroquine Synergistically Kill Glioma Cells by Inducing Organelle Stress and Disrupting Ca2+ Homeostasis. *Biochem Pharmacol*, 2020, *178*, 114098.

[41] Kumar, P.; Sharma, G.; Kumar, R.; Singh, B.; Malik, R.; Katare, O.P.; Raza, K. Promises of a Biocompatible Nanocarrier in Improved Brain Delivery of Quercetin: Biochemical, Pharmacokinetic and Biodistribution Evidences. *Int J Pharm*, 2016, *515*, 307–314.

[42] Wall, M.E.; Wani, M.C.; Cook, C.E.; Palmer, K.H.; McPhail, A.T.; Sim, G.A. Plant Antitumor Agents. I. The Isolation and Structure of Camptothecin, a Novel Alkaloidal Leukemia and Tumor Inhibitor from *Camptotheca Acuminata* 1,2. *J. Am. Chem. Soc.*, 1966, *88*, 3888–3890.

[43] Checa-Chavarria, E.; Rivero-Buceta, E.; Sanchez Martos, M.A.; Martinez Navarrete, G.; Soto-Sánchez, C.; Botella, P.; Fernández, E. Development of a Prodrug of Camptothecin for Enhanced Treatment of Glioblastoma Multiforme. *Mol Pharm*, 2021, *18*, 1558–1572.

[44] Taki, H.; Kanazawa, T.; Akiyama, F.; Takashima, Y.; Okada, H. Intranasal Delivery of Camptothecin-Loaded Tat-Modified Nanomicells for Treatment of Intracranial Brain Tumors. *Pharmaceuticals (Basel)*, 2012, *5*, 1092–1102.

[45] Ozeki, T.; Hashizawa, K.; Kaneko, D.; Imai, Y.; Okada, H. Treatment of Rat Brain Tumors Using Sustained-Release of Camptothecin from Poly(Lactic-Co-Glycolic Acid) Microspheres in a Thermoreversible Hydrogel. *Chem Pharm Bull (Tokyo)*, 2010, *58*, 1142–1147.

[46] Janss, A.J.; Levow, C.; Bernhard, E.J.; Muschel, R.J.; McKenna, W.G.; Sutton, L.; Phillips, P.C. Caffeine and Staurosporine Enhance the Cytotoxicity of Cisplatin and Camptothecin in Human Brain Tumor Cell Lines. *Exp Cell Res*, 1998, *243*, 29–38.

[47] Cırpanlı, Y.; Allard, E.; Passirani, C.; Bilensoy, E.; Lemaire, L.; Calış, S.; Benoit, J.-P. Antitumoral Activity of Camptothecin-Loaded Nanoparticles in 9L Rat Glioma Model. *Int J Pharm*, 2011, *403*, 201–206.

[48] Lin, C.-J.; Lin, Y.-L.; Luh, F.; Yen, Y.; Chen, R.-M. Preclinical Effects of CRLX101, an Investigational Camptothecin-Containing Nanoparticle Drug Conjugate, on Treating Glioblastoma Multiforme via Apoptosis and Antiangiogenesis. *Oncotarget*, 2016, *7*, 42408–42421.

[49] He, J.; Giusti, M.M. Anthocyanins: Natural Colorants with Health-Promoting Properties. *Annu Rev Food Sci Technol*, 2010, *1*, 163–187.

[50] Lamy, S.; Muhire, É.; Annabi, B. Antiproliferative Efficacy of Elderberries and Elderflowers (Sambucus Canadensis) on Glioma and Brain Endothelial Cells under Normoxic and Hypoxic Conditions. *J Funct Foods*, 2018, *40*, 164–179.

[51] Lamy, S.; Lafleur, R.; Bédard, V.; Moghrabi, A.; Barrette, S.; Gingras, D.; Béliveau, R. Anthocyanidins Inhibit Migration of Glioblastoma Cells: Structure-Activity Relationship and Involvement of the Plasminolytic System. *J Cell Biochem*, 2007, *100*, 100–111.

[52] Ouanouki, A.; Muhire, E.; Lamy, S.; Annabi, B. Abstract 4316: Pharmacological Targeting of the TGF-Beta-Induced Epithelial-Mesenchymal Transition by Anthocyanidins in Glioblastoma. *Cancer Research*, 2016, *76*, 4316.

[53] Ouanouki, A.; Lamy, S.; Annabi, B. Anthocyanidins Inhibit Epithelial-Mesenchymal Transition through a TGFβ/Smad2 Signaling Pathway in Glioblastoma Cells. *Mol Carcinog*, 2017, *56*, 1088–1099.

[54] Zhou, Y.; Chen, L.; Ding, D.; Li, Z.; Cheng, L.; You, Q.; Zhang, S. Cyanidin-3-O-Glucoside Inhibits the β-Catenin/MGMT Pathway by Upregulating MiR-214–5p to Reverse Chemotherapy Resistance in Glioma Cells. *Sci Rep*, 2022, *12*, 7773.

[55] Hsu, Y.-C.; Chen, M.-J.; Huang, T.-Y. Inducement of Mitosis Delay by Cucurbitacin E, a Novel Tetracyclic Triterpene from Climbing Stem of Cucumis Melo L., Through GADD45γ in Human Brain Malignant Glioma (GBM) 8401 Cells. *Cell Death Dis*, 2014, *5*, e1087.

[56] Cheng, A.-C.; Hsu, Y.-C.; Tsai, C.-C. The Effects of Cucurbitacin E on GADD45β-Trigger G2/M Arrest and JNK-Independent Pathway in Brain Cancer Cells. *J Cell Mol Med*, 2019, *23*, 3512–3519.

[57] Touihri-Barakati, I.; Kallech-Ziri, O.; Ayadi, W.; Kovacic, H.; Hanchi, B.; Hosni, K.; Luis, J. Cucurbitacin B Purified from Ecballium Elaterium (L.) A. Rich from Tunisia Inhibits A5β1 Integrin-Mediated Adhesion, Migration, Proliferation of Human Glioblastoma Cell Line and Angiogenesis. *Eur J Pharmacol*, 2017, *797*, 153–161.

[58] Yuan, G.; Yan, S.-F.; Xue, H.; Zhang, P.; Sun, J.-T.; Li, G. Cucurbitacin I Induces Protective Autophagy in Glioblastoma in Vitro and in Vivo. *J Biol Chem*, 2014, *289*, 10607–10619.

[59] Premkumar, D.R.; Jane, E.P.; Pollack, I.F. Cucurbitacin-I Inhibits Aurora Kinase A, Aurora Kinase B and Survivin, Induces Defects in Cell Cycle Progression and Promotes ABT-737-Induced Cell Death in a Caspase-Independent Manner in Malignant Human Glioma Cells. *Cancer Biol Ther*, 2015, *16*, 233–243.

Index

For Product Safety Concerns and Information please contact our EU
representative GPSR@taylorandfrancis.com
Taylor & Francis Verlag GmbH, Kaufingerstraße 24, 80331 München, Germany